MW01484279

Aviation Acronyms, Contractions, & Mnemonics

A complete pilot resource
of aviation terms for
every sector of aviation

Darren Smith, ATP, CFII/MEI
Editor

Darren Smith, ATP, CFII/MEI
Certificated Flight Instructor
www.cfidarren.com

*"...the safety of the operator is more important than any other point.
Greater prudence is needed rather than greater skill."*
— *Wilbur Wright, 1901*

Purpose of this guide: This guide is meant to offer pilots complete resource of aviation terms for every sector of aviation. As a review and reference for all pilots, it strives to present the information to keep you current. I would love to hear from you regarding your experience with this guide. Often pilots comment about the right way, the wrong way, and the FAA way. The result is most pilots chose the "practical way" which is a combination of all three. I caution all pilots to err on the side of the "safe way" so that you do not become a statistic.

Aviation Acronyms (PocketLearning) / Darren Smith
ISBN-13: 978-1468069501
ISBN-10: 1468069500

Table of Contents

UPDATES TO THIS BOOK ARE FREE

Please Register

www.cfidarren.com/register.htm

Protect your investment and register so that you can receive free publication updates, monthly newsletters, aviation safety notices, and free video training.

Metric Conversion Tables

Length and Distance

When you know:	Multiply by:	To find:
inches (in)	2.54	centimeters
feet (ft)	30.480	centimeters
yards	.9144	meters (m)
meters (m)	1.094	yards
statute miles	.8684	nautical miles
nautical miles	1.1516	statute miles
statute miles	1.609	kilometers
kilometers (km)	0.6213	statute miles
miles/hour	88	feet/minute

Volume and capacity (liquid)

When you know:	Multiply by:	To find:
gallons, US	3.785	liters
liters (L)	0.2642	gallons, US
gallons, Imp	4.546	liters
liters (L)	0.22	gallons, Imp

Weight and mass

When you know:	Multiply by:	To find:
ounces (oz)	28.350	grams
pounds (lb)	0.4536	kilograms
grams (g)	0.03527	ounces
kilograms	2.205	pounds

Temperature

When you know:	Formula:	To find:
°C	9/5°C+32	°F
°F	5/9(°F-32)	°C

Acronyms & Contractions Related to Winter & Icing

Runway Condition Reports

Devices which measure runway friction and braking action are abbreviated as follows:

BOW Bowmonk Decelerometer (Bowmonk Sales)
BRD Brakemeter-Dynometer
ERD Electronic Recording Decelerometer (Bowmonk)
GRT Griptester (Findlay, Irvine, LTD.)
MUM Mark 4 Mu Meter (Bison Instruments, Inc.)
RFT Runway friction tester (K.J. LAW Engineers, Inc.)
SFH Surface friction tester (high pressure tire)
        (Saab, Airport Surface Friction Tester AB)
SFL Surface friction tester (low pressure tire)
        (Saab, Airport Surface Friction Tester AB)
SKH Skiddometer (high pressure tire) (Airport Equipment Co.)
SKL Skiddometer (low pressure tire) (Airport Equipment Co.)
TAP Tapley Decelerometer (Tapley Sales)

Actual Runway Conditions Reported as Follows:
DRFT - Drifting or drifted snow
IR - Ice on runway
LSR - Loose snow on runway
PLW - Plowed/swept
PSR - Patches of snow on runway
PTCHY - Patchy (not completely covered)
RMNDR - surrounding area
SIR - Snow & Ice on runway
SLR - Slush on runway
SN - Snow
WSR - Wet snow on runway
WTR - Water on runway

*"One learns by doing a thing; for though you think
you know it, you have no certainty until you try."*
— *Sophocles*

Acronyms &
Contractions
Related to
NOTAMs

ABN   Airport Beacon
ABV   Above
ACC   Area Control Center (ARTCC)
ACCUM   Accumulate
ACFT   Aircraft
ACR   Air Carrier
ACT   Active
ADJ   Adjacent
ADZD   Advised
AFD   Airport Facility Directory
AGL   Above Ground Level
ALS   Approach Light System
ALT   Altitude
ALTM   Altimeter
ALTN   Alternate
ALTNLY   Alternately
ALSTG   Altimeter Setting
AMDT   Amendment
AMGR   Airport Manager
AMOS   Automatic Meteorological Observing System
AP   Airport
APCH   Approach
AP LGT   Airport Lights
APP   Approach Control
ARFF   Aircraft Rescue & Fire Fighting
ARR   Arrive, Arrival
ASOS   Automated Surface Observing System
ASPH   Asphalt
ATC   Air Traffic Control
ATCSCC   Air Traffic Control System Command Center
ATIS   Automatic Terminal Information Service
AUTH   Authority
AUTOB   Automatic Weather Reporting System
AVBL   Available
AWOS   Automatic Weather Observing/Reporting System
AWY   Airway
AZM   Azimuth
BA FAIR   Braking action fair
BA NIL   Braking action nil
BA POOR   Braking action poor

BC   Back Course
BCN   Beacon
BERM   Snowbank/s Containing Earth/Gravel
BLW   Below
BND   Bound
BRG   Bearing
BYD   Beyond
CAAS   Class A Airspace
CAT   Category
CBAS   Class B Airspace
CBSA   Class B Surface Area
CCAS   Class C Airspace
CCLKWS   Counterclockwise
CCSA   Class C Surface Area
CD   Clearance Delivery
CDAS   Class D Airspace
CDSA   Class D Surface Area
CEAS   Class E Airspace
CESA   Class E Surface Area
CFR   Code of Federal Regulations
CGAS   Class G Airspace
CHG   Change
CIG   Ceiling
CK   Check
CL   Center line
CLKWS   Clockwise
CLR   Clearance, Clear(s), Cleared To
CLSD   Closed
CMB   Climb
CMSND   Commissioned
CNL   Cancel
COM   Communications
CONC   Concrete
CPD   Coupled
CRS   Course
CTC   Contact
CTL   Control
DALGT   Daylight
DCMSND   Decommissioned
DCT   Direct

DEGS   Degrees
DEP   Depart/Departure
DEPPROC   Departure Procedures
DH   Decision Height
DISABLD   Disabled
DIST   Distance
DLA   Delay or Delayed
DLT   Delete
DLY   Daily
DME   Distance Measuring Equipment
DMSTN   Demonstration
DP   Dew Point Temperature
DRFT   Snowbank/s Caused By Wind Action
DSPLCD   Displaced
E   East
EB   Eastbound
EFAS   En Route Flight Advisory Service
ELEV   Elevation
ENG   Engine
ENRT   En route
ENTR   Entire
EXC   Except
FAC   Facility or Facilities
FAF   Final Approach fix
FAN MKR   Fan Marker
FDC   Flight Data Center
FI/T   Flight Inspection Temporary
FI/P   Flight Inspection Permanent
FM   From
FREQ   Frequency
FNA   Final Approach
FPM   Feet Per Minute
FREQ   Frequency
FRH   Fly Runway Heading
FRI   Friday
FRZN   Frozen
FSS   Automated/Flight Service Station
FT   Foot, feet
GC   Ground Control
GCA   Ground Control Approach

GOVT   Government
GP   Glide Path
GPS   Global Positioning System
GRVL   Gravel
HAA   Height Above Airport
HAT   Height Above Touchdown
HDG   Heading
HEL   Helicopter
HELI   Heliport
HIRL   High Intensity Runway Lights
HIWAS   Hazardous Inflight Weather Advisory Service
HLDG   Holding
HOL   Holiday
HP   Holding Pattern
HR   Hour
IAF   Initial approach fix
IAP   Instrument Approach Procedure
INBD   Inbound
ID   Identification
IDENT   Identify/Identifier/Identification
IF   Intermediate fix
ILS   Instrument Landing System
IM   Inner Marker
IMC   Instrument Meteorological Conditions
IN   Inch/Inches
INDEFLY   Indefinitely
INFO   Information
INOP   Inoperative
INSTR   Instrument
INT   Intersection
INTL   International
INTST   Intensity
IR   Ice On Runway/s
KT   Knots
L   Left
LAA   Local Airport Advisory
LAT   Latitude
LAWRS   Limited Aviation Weather Reporting Station
LB   Pound/Pounds
LC   Local Control

LOC   Local/Locally/Location
LCTD   Located
LDA   Localizer Type Directional Aid
LGT   Light or Lighting
LGTD   Lighted
LIRL   Low Intensity Runway Lights
LLWAS   Low Level Wind Shear Alert System
LM   Compass Locator at ILS Middle Marker
LDG   Landing
LLZ   Localizer
LO   Compass Locator at ILS Outer Marker
LONG   Longitude
LRN   Loran
LSR   Loose Snow on Runway/s
LT   Left Turn
MAG   Magnetic
MAINT   Maintain, maintenance
MALS   Medium Intensity Approach Light System
MALSF   Medium Intensity Approach Light System with Sequenced Flashers
MALSR   Medium Intensity Approach Light System with Runway Alignment Indicator Lights
MAPT   Missed Approach Point
MCA   Minimum Crossing Altitude
MDA   Minimum Descent Altitude
MEA   Minimum En Route Altitude
MED   Medium
MIN   Minute
MIRL   Medium Intensity Runway Lights
MLS   Microwave Landing System
MM   Middle Marker
MNM   Minimum
MNT   Monitor/Monitoring/Monitored
MOC   Minimum Obstruction Clearance
MON   Monday
MRA   Minimum reception altitude
MSA   Minimum Safe Altitude/Minimum Sector Altitude
MSAW   Minimum Safe Altitude Warning
MSG   Message
MSL   Mean Sea Level

MU   MU Meters
MUD   Mud
MUNI   Municipal
N   North
NA   Not Authorized
NAV   Navigation
NB   Northbound
NDB   Nondirectional Radio Beacon
NE   Northeast
NGT   Night
NM   Nautical Mile(s)
NMR   Nautical Mile Radius
NONSTD   Nonstandard
NOPT   No Procedure Turn Required
NR   Number
NTAP   Notice To Airmen Publication
NW   Northwest
OBSC   Obscured
OBST   Obstruction
OM   Outer Marker
OPR   Operate
OPS   Operation
ORIG   Original
OTS   Out of Service
OVR   Over
PAEW   Personnel and Equipment Working
PAPI   Precision Approach Path Indicator
PAR   Precision Approach Radar
PARL   Parallel
PAT   Pattern
PAX   Passenger
PCL   Pilot Controlled Lighting
PERM   Permanent/Permanently
PJE   Parachute Jumping Exercise
PLA   Practice Low Approach
PLW   Plow/Plowed
PN   Prior Notice Required
PPR   Prior Permission Required
PREV   Previous
PRN   Psuedo random noise

PROC   Procedure
PROP   Propeller
PSR   Packed Snow on Runway/s
PTCHY   Patchy
PTN   Procedure Turn
PVT   Private
RAIL   Runway Alignment Indicator Lights
RAMOS   Remote Automatic Meteorological Observing System
RCAG   Remote Communication Air/Ground Facility
RCL   Runway Centerline
RCLL   Runway Centerline Light System
RCO   Remote Communication Outlet
REC   Receive/Receiver
RELCTD   Relocated
REIL   Runway End Identifier Lights
REP   Report
RLLS   Runway Lead-in Lights System
RMNDR   Remainder
RNAV   Area Navigation
RPLC   Replace
RQRD   Required
RRL   Runway Remaining Lights
RSR   En route Surveillance Radar
RSVN   Reservation
RT   Right Turn
RTE   Route
RTR   Remote Transmitter/Receiver
RTS   Return to Service
RUF   Rough
RVR   Runway Visual Range
RVRM   Runway Visual Range Midpoint
RVRR   Runway Visual Range Rollout
RVRT   Runway Visual Range Touchdown
RWY   Runway
S   South
SA   Sand, sanded
SAT   Saturday
SAWR   Supplementary Aviation Weather Reporting Station
SB   Southbound

SDF   Simplified Directional Facility
SE   Southeast
SFL   Sequence Flashing Lights
SID   Standard Instrument Departure
SIMUL   Simultaneous
SIR   Packed or Compacted Snow and Ice on Runway/s
SKED   Scheduled
SLR   Slush on Runway/s
SN   Snow
SNBNK   Snowbank/s Caused by Plowing
SNGL   Single
SPD   Speed
SSALF   Simplified Short Approach Lighting System with Sequenced Flashers
SSALR   Simplified Short Approach Lighting System with Runway Alignment Indicator Lights
SSALS   Simplified Short Approach Lighting System
SSR   Secondary Surveillance Radar
STA   Straight-in Approach
STAR   Standard Terminal Arrival
SUN   Sunday
SVC   Service
SW   Southwest
SWEPT   Swept or Broom/Broomed
T   Temperature
TAA   Terminal Arrival Area
TACAN   Tactical Air Navigational Aid
TAR   Terminal Area Surveillance Radar
TDZ   Touchdown Zone
TDZ LG   Touchdown Zone Lights
TEMPO   Temporary
TFC   Traffic
TFR   Temporary Flight Restriction
TGL   Touch and Go Landings
THN   Thin
THR   Threshold
THRU   Through
THU   Thursday
TIL   Until
TKOF   Takeoff

TM   Traffic Management
TMPA   Traffic Management Program Alert
TRML   Terminal
TRNG   Training
TRSN   Transition
TSNT   Transient
TUE   Tuesday
TWR   Tower
TWY   Taxiway
UFN   Until Further Notice
UNAVBL   Unavailable
UNLGTD   Unlighted
UNMKD   Unmarked
UNMNT   Unmonitored
UNREL   Unreliable
UNUSBL   Unusable
VASI   Visual Approach Slope Indicator
VDP   Visual Descent Point
VGSI   Visual Glide Slope Indicator
VIA   By Way Of
VICE   Instead/Versus
VIS   Visibility
VMC   Visual Meteorological Conditions
VOL   Volume
VOR   VHF Omni-Directional Radio Range
VORTAC   VOR and TACAN (colocated)
W   West
WB   Westbound
WED   Wednesday
WEF   With Effect From or Effective From
WI   Within
WIE   With Immediate Effect or Effective Immediately
WKDAYS   Monday through Friday
WKEND   Saturday and Sunday
WND   Wind
WPT   Waypoint
WSR   Wet Snow on Runway/s
WTR   Water on Runway/s
WX   Weather

Acronyms & Contractions Related to METARs

METAR Weather Contractions

Weather Identifiers:

B - Began
BC Patches
BL Blowing
BR Mist
DR Low Drifting
DS Dust storm
DU Dust
DZ Drizzle
E - Ended

FC, +FC Funnel Cloud, Well-Developed Funnel Cloud, Tornado or Waterspout
FG Fog
FU Smoke
FZ Freezing
GR Hail
GS Small Hail or Snow Pellets (< 1/4)

HZ Haze
IC Ice Crystals
MI Shallow
PE Ice Pellets
PO Well-Developed Dust/Sand Whirls
PR Partial
PY Spray
RA Rain
SA Sand

SG Snow Grains
SH Showers
SN Snow
SQ Squalls Moderate
SS Sandstorm
TS Thunderstorm
UP Unknown Precip
VA Volcanic Ash
VC In the Vicinity

Modifiers: - Light   + Heavy   P More than   M Less than   B Began   E Ended

Sky Conditions:

BKN – Broken cloud layer 5/8ths to 7/8ths
CB – Cumulonimbus
CLR – Sky clear at or below 12,000AGL
FEW – Few cloud layer 0/8ths to 2/8ths

OVC – Overcast cloud layer 8/8ths coverage
SCT – Scattered cloud layer 3/8ths to 4/8ths
SKC – Sky Clear
TCU – Towering Cumulus

Other:

A01 – Automated Observation without precipitation discrimination
A02 – Automated Observation with precipitation discrimination
A3000 – Altimeter setting 30.00"
AMD – Amended forecast
AUTO – without human editing
BECMG – Becoming…
BECMG 0002 = becoming 00 to 02 Zulu
CAVU – Ceiling and visibility unlimited
COR – Correction
DSNT – Distant weather phenomenon
FM – From…   FM0200 = from 0200 Zulu
FROPA – Frontal Passage
LDG – Landing
M – Minus, below zero, "less than"
NO – Not available
NSW – No significant weather
P6SM – Plus 6 Statute Miles, greater than, "more than"
PK WND – Peak Wind
PRESFR – Pressure Falling Rapidly
PRESRR – Pressure Rising Rapidly

PROB40 – Probability of 40%
R04 – Runway 4
RMK – Remarks
RWY – Runway
RVRNO – Runway Visual Range not available
SFC VIS – Surface Visibility
SLP – Sea Level Pressure, add 10 to beginning of numbers given to make XXXX.x
SLPNO – Sea Level Pressure not available
SM – Statute miles
SPECI – Special Report
TEMPO – Temporarily…
TEMPO 0002 = Temporarily 00 to 02 Zulu
T02560179 – Temperature 25.6 dew point 17.9
TWR VIS – Tower Visibility
V – Varying
VRB – Variable
VRB VIS – Variable Visibility
VV – Vertical Visibility, indefinite ceiling
WS – Wind shear
WSHFT - Wind shift

Acronyms & Contractions Related to Weather Reports

AAF - Army Air Field
AAL - above aerodrome level
AATM - at all times
ABD – aboard
ABNDT - abundant
ABNML - abnormal
ABT - about
ABV - above
AC - altocumulus
ACARS - Aircraft communication addressing & reporting system
ACAS - airborne collision avoidance system
ACCUM - accumulate
ACFT - aircraft
ACK - acknowledge
ACL - altimeter check location
ACLD - above clouds
ACLT - accelerate
ACPT - accept
ACR - air carrier
ACRBT - acrobatic
ACRS - across
ACSL - standing lenticular altocumulus
ACT - active or activated or activity
ACTG - acting
ACTV - active
ACTVT - activate
ACYC - anticyclonic
ADA - advisory area
ADDN - addition
ADF - automatic direction finder
ADIZ - air defense identification zone
ADJ - adjacent
ADQT - adequate
ADRNDCK - Adirondack
ADV - advise
ADVCTN - advection
ADVN - advance
ADVY - advisory
ADVZY - advisory

AFB - Air Force Base
AFCT - affect
AFD - Airport/Facility Directory
AFDK - after dark
AFSS - Automated Flight Service Station
AFT - after
AFTN - afternoon
AGL - above ground level
AHD - ahead
AIM - Aeronautical Information Manual
AIRMET - Airmen's Meteorological Information
ALF - aloft
ALG - along
ALGHNY - Allegheny
ALNMT - alignment
ALQDS - all quadrants
ALS - approach light system
ALSEC - all sectors
ALSF-1 - standard 2400' high-intensity approach lighting system with sequenced flashers (Category I configuration)
ALSF-2 - standard 2400' high-intensity approach lighting system with sequenced flashers (Category II configuration)
ALSTG - altimeter setting
ALT - altitude
ALTA - Alberta
ALTM - altimeter
ALTN - alternate
ALUTN - Aleutian
AM - ante meridian
AMS - air mass
AMSL - above mean sea level
ANCPT - anticipate
ANLYS - analysis
AO1 - ASOS automated observation without precipitation discriminator (rain/

snow)

AO2 - ASOS automated observation with precipitation discriminator (rain/snow)

AOA - at or above

AOB - at or below

AOE - airport of entry

APCH - approach

APL - airport lights

APLCN - Appalachian

APN - Apron

APRNT - apparent

APROP - appropriate

APRX - approximate

ARB - Air Reserve Base

ARFOR - area forecast

ARINC - Aeronautical Radio, Incorporated

ARND - around

ARTC - air route traffic control

ARTCC - Air Route Traffic Control Center

AS - altostratus

ASAP - as soon as possible

ASDA - accelerate-stop distance available

ASL - above sea level

ASOS - automated surface observing system

ASPH - asphalt

ATA - actual time of arrival

ATIS - automatic terminal information service

ATLC - Atlantic

AURBO - Aurora Borealis

AUTH - authorized

AUTO - automatic

AVG - average

AVGAS - aviation gasoline

AWOS - automatic weather observing/reporting system

AWY - airway

AZM – azimuth

BACLIN - Baroclinic

BAJA - Baja California

BATROP - Barotropic

BC - British Columbia

BCH - Beach

BCKG - Backing

BCM - Become

BCMG - Becoming

BCMS - Becomes

BD - Blowing dust

BDA - Bermuda

BDRY - Boundary

BFDK - Before dark

BFR - Before

BGN - Begin

BGNG - Beginning

BGNS - Begins

BHND - Behind

BINOVC - Breaks in overcast

BKN - Broken

BLD - Build

BLDG - Building

BLDS - Builds

BLDUP - Buildup

BLKHLS - Black Hills

BLKT - Blanket

BLKTG - Blanketing

BLKTS - Blankets

BLO - Below

BLZD - Blizzard

BN - Blowing sand

BND - Bound

BNDRY - Boundary

BNDRYS - Boundaries

BNTH - Beneath

BOOTHEEL - Bootheel

BR - Branch

BRG - Branching

BRS - Branches

BRF - Brief

BRK - Break

BRKG - Breaking
BRKHIC - Breaks in higher clouds
BRKS - Breaks
BRKSHR - Berkshire
BRKSHRS - Berkshires
BRM - Barometer
BS - Blowing snow
BTWN - Between
BYD - Beyond
C - Celsius
CA - California
CAA - Cold Air Advection
CARIB - Caribbean
CASCDS - Cascades
CAVOK - Ceiling and visibility OK
CAVU - Ceiling and visibility unlimited
CB - Cumulonimbus
CBS - Cumulonimbi
CC - Cirrocumulus
CCLDS - Clear of clouds
CCLKWS - Counter-clockwise
CCSL - Standing lenticular cirrocumulus
CDFNT - Cold front
CDFNTL - Cold frontal
CFP - Cold front passage
CG - Cloud-to-ground
CHC - Chance
CHCS - Chances
CHG - Change
CHGD - Changed
CHGG - Changing
CHGS - Changes
CHSPK - Chesapeake
CI - Cirrus
CIG - Ceiling
CIGS - Ceilings
CLD - Cloud
CLDNS - Cloudiness
CLDS - Clouds
CLKWS - Clockwise
CLR - Clear

CLRG - Clearing
CLRS - Clears
CMPLX - Complex
CNCL - Cancel
CNCLD - Cancelled
CNCLG - Canceling
CNCLS - Cancels
CNDN - Canadian
CNTR - Center
CNTRD - Centered
CNTRLN - Centerline
CNTRS - Centers
CNTRL - Central
CNTY - County
CNTYS - Counties
CNVG - Converge
CNVGG - Converging
CNVGNC - Convergence
CNVTN - Convection
CNVTV - Convective
CNVTVLY - Convectively
CNFDC - Confidence
CO - Colorado
COMPAR - Compare
COMPARG - Comparing
COMPARD - Compared
COMPARS - Compares
COND - Conditions
CONT - Continue
CONTD - Continued
CONTLY - Continually
CONTG - Continuing
CONTRAILS - Condensation Trails
CONTS - Continues
CONTDVD - Continental Divide
CONUS - Continental U.S.
COORD - Coordinate
COR - Correction
CPBL - Capable
CRC - Circle
CRCLC - Circulate

CRCLN - Circulation
CRNR - Corner
CRNRS - Corners
CRS - Course
CS - Cirrostratus
CSDR - Consider
CSDRBL - Considerable
CST - Coast
CSTL - Coastal
CT - Connecticut
CTGY - Category
CTSKLS - Catskills
CU - Cumulus
CUFRA - Cumulus Fractus
CVR - Cover
CVRD - Covered
CVRG - Covering
CVRS - Covers
CYC - Cyclonic
CYCLGN - Cyclogenesis
DABRK - Daybreak
DALGT - Day light
DBL - Double
DC - District of Columbia
DCR - Decrease
DCRD - Decreased
DCRG - Decreasing
DCRGLY - Decreasingly
DCRS - Decreases
DE - Delaware
DEG - Degree
DEGS - Degrees
DELMARVA - Delaware-Maryland-Virginia
DFCLT - Difficult
DFCLTY - Difficulty
DFNT - Definite
DFNTLY - Definitely
DFRS - Differs
DFUS - Diffuse
DGNL - Diagonal

DGNLLY - Diagonally
DIGG - Digging
DIR - Direction
DISC - Discontinue
DISCD - Discontinued
DISCG - Discontinuing
DISRE - Disregard
DISRED - Disregarded
DISREG - Disregarding
DKTS - Dakotas
DLA - Delay
DLAD - Delayed
DLT - Delete
DLTD - Deleted
DLTG - Deleting
DLY - Daily
DMG - Damage
DMGD - Damaged
DMGG - Damaging
DMNT - Dominant
DMSH - Diminish
DMSHD - Diminished
DMSHG - Diminishing
DMSHS - Diminishes
DNDFTS - Downdrafts
DNS - Dense
DNSLP - Downslope
DNSTRM - Downstream
DNWND - Down wind
DP - Deep
DPND - Deepened
DPNS - Deepens
DPR - Deeper
DPNG - Deepening
DPTH - Depth
DRFT - Drift
DRFTD - Drifted
DRFTG - Drifting
DRFTS - Drifts
DRZL - Drizzle
DSCNT - Descent

DSIPT - Dissipate
DSIPTD - Dissipated
DSIPTG - Dissipating
DSIPTN - Dissipation
DSIPTS - Dissipates
DSND - Descend
DSNDG - Descending
DSNDS - Descends
DSNT - Distant
DSTBLZ - Destabilize
DSTBLZD - Destabilized
DSTBLZG - Destabilizing
DSTBLZS - Destabilizes
DSTBLZN - Destabilization
DSTC - Distance
DTRT - Deteriorate
DTRTD - Deteriorated
DTRTG - Deteriorating
DTRTS - Deteriorates
DURG - During
DURN - Duration
DVLP - Develop
DVLPD - Developed
DVLPG - Developing
DVLPMT - Development
DVLPS - Develops
DVRG - Diverge
DVRGG - Diverging
DVRGNC - Divergence
DVRGS - Diverges
DVV - Downward vertical velocity
DWNDFTS - Downdrafts
DWPNT - Dewpoint
DWPNTS - Dewpoints
DX - Duplex
E - East
EBND - East bound
EFCT - Effect
ELNGT - Elongate
ELNGTD - Elongated
ELSW - Elsewhere

EMBDD - Embedded
EMERG - Emergency
ENCTR - Encounter
ENDG - Ending
ENE - East-northeast
ENELY - East-northeasterly
ENERN - East-northeastern
ENEWD - East-northeastward
ENHNC - Enhance
ENHNCD - Enhanced
ENHNCG - Enhancing
ENHNCS - Enhances
ENHNCMNT - Enhancement
ENTR - Entire
ERN - Eastern
ERY - Early
ERYR - Earlier
ESE - East-southeast
ESELY - East-southeasterly
ESERN - East-southeastern
ESEWD - East-southeastward
ESNTL - Essential
ESTAB - Establish
ESTS - Estimates
ETA - Est time of arrival
ETC - Et cetera
ETIM - Elapsed time
EVE - Evening
EWD - Eastward
EXCLV - Exclusive
EXCLVLY - Exclusively
EXCP - Except
EXPC - Expect
EXPCD - Expected
EXPCG - Expecting
EXTD - Extend
EXTDD - Extended
EXTDG - Extending
EXTDS - Extends
EXTN - Extension
EXTRAP - Extrapolate

EXTRAPD - Extrapolated
EXTRM - Extreme
EXTRMLY - Extremely
EXTSV - Extensive
F - Fahrenheit
FA - Aviation area forecast
FAH - Fahrenheit
FAM - Familiar
FCST - Forecast
FCSTD - Forecasted
FCSTG - Forecasting
FCSTR - Forecaster
FCSTS - Forecasts
FIG - Figure
FILG - Filling
FIRAV - First available
FL - Florida
FLG - Falling
FLRY - Flurry
FLRYS - Flurries
FLT - Flight
FLW - Follow
FLWG - Following
FM - From
FMT - Format
FNCTN - Function
FNT - Front
FNTL - Frontal
FNTS - Fronts
FNTGNS - Frontogenesis
FNTLYS - Frontolysis
FORNN - Forenoon
FPM - Feet per minute
FQT - Frequent
FQTLY - Frequently
FRM - Form
FRMG - Forming
FRMN - Formation
FROPA - Frontal passage
FROSFC - Frontal surface
FRST - Frost

FRWF - Forecast wind factor
FRZ - Freeze
FRZN - Frozen
FRZG - Freezing
FT - Feet
FT - Terminal forecast
FTHR - Further
FVRBL - Favorable
FWD - Forward
FYI - For your information
G - Gust
GA - Georgia
GEN - General
GENLY - Generally
GEO - Geographic
GEOREF - Geographical reference
GF - Ground fog
GICG - Glaze icing
GLFALSK - Gulf of Alaska
GLFCAL - Gulf of California
GLFMEX - Gulf of Mexico
GLFSTLAWR - Gulf of St. Lawrence
GND - Ground
GNDFG - Ground fog
GRAD - Gradient
GRDL - Gradual
GRDLY - Gradually
GRT - Great
GRTLY - Greatly
GRTR - Greater
GRTST - Greatest
GRTLKS - Great Lakes
GSTS - Gusts
GSTY - Gusty
GV - Ground visibility
HAZ - Hazard
HCVIS - High clouds visible
HDFRZ - Hard freeze
HDSVLY - Hudson Valley
HDWND - Head wind
HGT - Height

HI - High
HIER - Higher
HIFOR - High level forecast
HLF - Half
HLTP - Hilltop
HLSTO - Hailstones
HLYR - Haze layer aloft
HND - Hundred
HR - Hour
HRS - Hours
HRZN - Horizon
HTG - Heating
HURCN - Hurricane
HUREP - Hurricane report
HV - Have
HVY - Heavy
HVYR - Heavier
HVYST - Heaviest
HWVR - However
HWY - Highway
IA - Iowa
IC - Ice
ICG - Icing
ICGIC - Icing in clouds
ICGIP - Icing in precipitation
ID - Idaho
IL - Illinois
IMDT - Immediate
IMDTLY - Immediately
IMPL - Impulse
IMPLS - Impulses
IMPT - Important
INCL - Include
INCLD - Included
INCLG - Including
INCLS - Includes
INCR - Increase
INCRD - Increased
INCRG - Increasing
INCRGLY - Increasingly
INCRS - Increases

INDC - Indicate
INDCD - Indicated
INDCG - Indicating
INDCS - Indicates
INDEF - Indefinite
INFO - Information
INLD - Inland
INSTBY - Instability
INTCNTL - Intercontinental
INTL - International
INTMD - Intermediate
INTMT - Intermittent
INTMTLY - Intermittently
INTR - Interior
INTRMTRGN - Intermountain region
INTS - Intense
INTSFCN - Intensification
INTSFY - Intensify
INTSFYD - Intensified
INTSFYG - Intensifying
INTSFYS - Intensifies
INTSTY - Intensity
INTVL - Interval
INVRN - Inversion
IOVC - In overcast
INVOF - In vicinity of
IP - Ice pellets
IPV - Improve
IPVG - Improving
IR - Infrared
ISOL - Isolate
ISOLD - Isolated
JCTN - Junction
JTSTR - Jet stream
KFRST - Killing frost
KLYR - Smoke layer aloft
KOCTY - Smoke over city
KS - Kansas
KT - Knots
KY - Kentucky
LA - Louisiana

LABRDR - Labrador
LAT - Latitude
LCL - Local
LCLY - Locally
LCTD - Located
LCTN - Location
LCTMP - Little change in temperature
LEVEL - Level
LFM - Limited Fine Mesh Model
LFTG - Lifting
LGRNG - Long range
LGT - Light
LGTR - Lighter
LGWV - Long wave
LI - Lifted Index
LIS - Lifted indices
LK - Lake
LKS - Lakes
LKLY - Likely
LLJ - Low Level Jet
LLWS - Low Level Wind Shear
LLWAS - Low level wind shear alert system
LMTD - Limited
LMTG - Limiting
LMTS - Limits
LN - Line
LN - Lines
LO - Low
LONG - Longitude
LONGL - Longitudinal
LRG - Large
LRGLY - Largely
LRGR - Larger
LRGST - Largest
LST - Local standard time
LTD - Limited
LTG - Lightning
LTGCC - Lightning cloud-to-cloud
LTGCG - Lightning cloud-to-ground
LTGCCCG - Lightning cloud-to-cloud

cloud-to-ground
LTGCW - Lightning cloud-to-water
LTGIC - Lightning in cloud
LTL - Little
LTLCG - Little change
LTR - Later
LTST - Latest
LV - Leaving
LVL - Level
LVLS - Levels
LWR - Lower
LWRD - Lowered
LWRG - Lowering
LYR - Layer
LYRD - Layered
LYRS - Layers
MA - Massachusetts
MAN - Manitoba
MAX - Maximum
MB - Millibars
MCD - Mesoscale discussion
MD - Maryland
MDFY - Modify
MDFYD - Modified
MDFYG - Modifying
MDL - Model
MDLS - Models
MDT - Moderate
MDTLY - Moderately
ME - Maine
MED - Medium
MEGG - Merging
MESO - Mesoscale
MET - Meteorological
METRO - Metropolitan
MEX - Mexico
MHKVLY - Mohawk Valley
MI - Michigan
MID - Middle
MIDN - Midnight
MIL - Military

MIN - Minimum
MISG - Missing
MLTLVL - Melting level
MN - Minnesota
MNLND - Mainland
MNLY - Mainly
MO - Missouri
MOGR - Moderate or greater
MOV - Move
MOVD - Moved
MOVG - Moving
MOVMT - Movement
MOVS - Moves
MPH - Miles per hour
MRGL - Marginal
MRGLLY - Marginally
MRNG - Morning
MRTM - Maritime
MS - Mississippi
MSG - Message
MSL - Mean sea level
MST - Most
MSTLY - Mostly
MSTR - Moisture
MT - Montana
MTN - Mountain
MTNS - Mountains
MULT - Multiple
MULTILVL - Multi-level
MXD - Mixed
N - North
NAB - Not above
NAT - North Atlantic
NATL - National
NAV - Navigation
NB - New Brunswick
NBND - Northbound
NBRHD - Neighborhood
NC - North Carolina
NCWX - No change in weather
ND - North Dakota

NE - Northeast
NEB - Nebraska
NEC - Necessary
NEG - Negative
NEGLY - Negatively
NELY - Northeasterly
NERN - Northeastern
NEWD - Northeastward
NEW ENG - New England
NFLD - Newfoundland
NGM - Nested Grid Model
NGT - Night
NH - New Hampshire
NIL - None
NJ - New Jersey
NL - No layers
NLT - Not later than
NLY - Northerly
NM - New Mexico
NMBR - Number
NMBRS - Numbers
NMC - National Meteorological Center
NML - Normal
NMRS - Numerous
NNE - North-northeast
NNELY - North-northeasterly
NNERN - North-northeastern
NNEWD - North-northeastward
NNW - North-northwest
NNWLY - North-northwesterly
NNWRN - North-northwestern
NNWWD - North-northwestward
NNNN - End of message
NOAA - National Oceanic and Atmospheric Administration
NOPAC - Northern Pacific
NPRS - Nonpersistent
NR - Near
NRLY - Nearly
NRN - Northern
NRW - Narrow

NS - Nova Scotia
NTFY - Notify
NTFYD - Notified
NV - Nevada
NVA - Negative vorticity advection
NW - Northwest
NWD - Northward
NWLY - Northwesterly
NWRN - Northwestern
NWS - National Weather Service
NY - New York
NXT - Next
OAT - Outside Air Temperature OBND - Outbound
OBS - Observation
OBSC - Obscure
OBSCD - Obscured
OBSCG - Obscuring
OCFNT - Occluded front
OCLD - Occlude
OCLDS - Occludes
OCLDD - Occluded
OCLDG - Occluding
OCLN - Occlusion
OCNL - Occasional
OCNLY - Occasionally
OCR - Occur
OCRD - Occurred
OCRG - Occurring
OCRS - Occurs
OFC - Office
OFP - Occluded frontal passage
OFSHR - Offshore
OH - Ohio
OK - Oklahoma
OMTNS - Over mountains
ONSHR - On shore
OR - Oregon
ORGPHC - Orographic
ORIG - Original
OSV - Ocean station vessel

OTLK - Outlook
OTP - On top
OTR - Other
OTRW - Otherwise
OUTFLO - Outflow
OVC - Overcast
OVNGT - Overnight
OVR - Over
OVRN - Overrun
OVRNG - Overrunning
OVTK - Overtake
OVTKG - Overtaking
OVTKS - Overtakes
PA - Pennsylvania
PAC - Pacific
PBL - Planetary boundary layer
PCPN - Precipitation
PD - Period
PDS - Periods
PDMT - Predominant
PEN - Peninsula
PERM - Permanent
PGTSND - Puget Sound
PHYS - Physical
PIBAL - Pilot balloon observation
PIBALS - Pilot balloon reports
PIREP - Pilot weather report
PIREPS - Pilot weather reports
PLNS - Plains
PLS - Please
PLTO - Plateau
PM - Post meridian
PNHDL - Panhandle
POS - Positive
POSLY - Positively
PPINE - PPI no echoes
PPSN - Present position
PRBL - Probable
PRBLY - Probably
PRBLTY - Probability
PRECD - Precede

PRECDD - Preceded
PRECDG - Preceding
PRECDS - Precedes
PRES - Pressure
PRESFR - Pressure falling rapidly
PRESRR - Pressure rising rapidly
PRIM - Primary
PRIN - Principal
PRIND - Present indications are
PRJMP - Pressure jump
PROC - Procedure
PROD - Produce
PRODG - Producing
PROG - Forecast
PROGD - Forecasted
PROGS - Forecasts
PRSNT - Present
PRSNTLY - Presently
PRST - Persist
PRSTS - Persists
PRSTNC - Persistence
PRSTNT - Persistent
PRVD - Provide
PRVDD - Provided
PRVDG - Providing
PRVDS - Provides
PS - Plus
PSBL - Possible
PSBLY - Possibly
PSBLTY - Possibility
PSG - Passage
PSN - Position
PSND - Positioned
PTCHY - Patchy
PTLY - Partly
PTNL - Potential
PTNLY - Potentially
PTNS - Portions
PUGET - Puget Sound
PVA - Positive vorticity advection
PVL - Prevail

PVLD - Prevailed
PVLG - Prevailing
PVLS - Prevails
PVLT - Prevalent
PWR - Power
QN - Question
QPFERD - NMC excessive rainfall discussion
QPFHSD - NMC heavy snow discussion
QPFSPD - NMC special precipitation discussion
QSTNRY - Quasistationary
QUAD - Quadrant
QUE - Quebec
R - Rain
RADAT - Radiosonde observation data
RAOB - Radiosonde observation
RAOBS - Radiosonde observations
RCH - Reach
RCHD - Reached
RCHG - Reaching
RCHS - Reaches
RCKY - Rocky
RCKYS - Rockies
RCMD - Recommend
RCMDD - Recommended
RCMDG - Recommending
RCMDS - Recommends
RCRD - Record
RCRDS - Records
RCV - Receive
RCVD - Received
RCVG - Receiving
RCVS - Receives
RDC - Reduce
RDGG - Ridging
RDR - Radar
RDVLP - Redevelop
RDVLPG - Redeveloping

RDVLPMT - Redevelopment
RE - Regard
RECON - Reconnaissance
REF - Reference
RES - Reserve
RGL - Regional Model
RGT - Right
RHINO - RHI not operative
RI - Rhode Island
RLBL - Reliable
REPL - Replace
REPLD - Replaced
REPLG - Replacing
REPLS - Replaces
REQ - Request
REQS - Requests
REQSTD - Requested
RESP - Response
RESTR - Restrict
RGD - Ragged
RGLR - Regular
RGN - Region
RGNS - Regions
RH - Relative Humidity
RIOGD - Rio Grande
RLTV - Relative
RLTVLY - Relatively
RMN - Remain
RMND - Remained
RMNDR - Remainder
RMNG - Remaining
RMNS - Remains
RNFL - Rainfall
ROT - Rotate
ROTD - Rotated
ROTG - Rotating
ROTS - Rotates
RPD - Rapid
RPDLY - Rapidly
RPLC - Replace
RPLCD - Replaced

RPLCG - Replacing
RPLCS - Replaces
RPRT - Report
RPRTD - Reported
RPRTG - Reporting
RPRTS - Reports
RPT - Repeat
RPTG - Repeating
RPTS - Repeats
RQR - Require
RQRD - Required
RQRG - Requiring
RQRS - Requires
RS - Receiver station
RSG - Rising
RSN - Reason
RSNG - Reasoning
RSNS - Reasons
RSTR - Restrict
RSTRD - Restricted
RSTRG - Restricting
RSTRS - Restricts
RTRN - Return
RTRND - Returned
RTRNG - Returning
RTRNS - Returns
RUF - Rough
RUFLY - Roughly
RVS - Revise
RVSD - Revised
RVSG - Revising
RVSS - Revises
RW - Rain shower
S - South
SA - Surface observation
SAO - Surface observation
SAOS - Surface observations
SASK - Saskatchewan
SATFY - Satisfactory
SBND - South bound
SBSD - Subside

SBSDD - Subsided
SBSDNC - Subsidence
SBSDS - Subsides
SC - South Carolina
SCND - Second
SCNDRY - Secondary
SCSL - Standing lenticular stratocumulus
SCT - Scatter
SCTD - Scattered
SCTR - Sector
SD - South Dakota
SE - Southeast
SEC - Second
SELS - Severe Local Storms Unit
SELY - Southeasterly
SEPN - Separation
SEQ - Sequence
SERN - Southeastern
SEWD -Southeastward
SFC - Surface
SFERICS - Atmospherics
SG - Snow grains
SGFNT - Significant
SGFNTLY - Significantly
SHFT - Shift
SHFTD - Shifted
SHFTG - Shifting
SHFTS - Shifts
SHLD - Shield
SHLW - Shallow
SHRT - Short
SHRTLY - Shortly
SHRTWV - Shortwave
SHRTWVS - Shortwaves
SHUD - Should
SHWR - Shower
SHWRS - Showers
SIERNEV - Sierra Nevada
SIG - Signature
SIGMET - Significant meteorological information

SIMUL - Simultaneous
SKC - Sky clear
SKED - Schedule
SLD - Solid
SLGT - Slight
SLGTLY - Slightly
SLO - Slow
SLOLY - Slowly
SLOR - Slower
SLP - Slope
SLPG - Sloping
SLT - Sleet
SLY - Southerly
SM - Statute mile
SMK - Smoke
SML - Small
SMLR - Smaller
SMRY - Summary
SMS - Synchronous meteorological satellite
SMTH - Smooth
SMTHR - Smoother
SMTHST - Smoothest
SMTM - Sometime
SMWHT - Somewhat
SNBNK - Snow bank
SND - Sand
SNFLK - Snow flake
SNGL - Single
SNOINCR - Snow increase
SNOINCRG - Snow increasing
SNST - Sunset
SNW - Snow
SNWFL - Snowfall
SOP - Standard operating procedure
SP - Snow pellets
SPCLY - Especially
SPD - Speed
SPDS - Speeds
SPENES - Satellite precipitation estimate statement

SPKL - Sprinkle
SPKLS - Sprinkles
SPLNS - Southern Plains
SPRD - Spread
SPRDG - Spreading
SPRDS - Spreads
SPRL - Spiral
SQAL - Squall
SQLN - Squall line
SR - Sunrise
SRN - Southern
SRND - Surround
SRNDD - Surrounded
SRNDG - Surrounding
SRNDS - Surrounds
SS - Sunset
SSE - South-southeast
SSELY - South-southeasterly
SSERN - South-southeastern
SSEWD - South-southeastward
SSW - South-southwest
SSWLY - South-southwesterly
SSWRN - South-southwestern
SSWWD - South-southwestward
ST - Stratus
STAGN - Stagnation
STBL - Stable
STBLTY - Stability
STD - Standard
STDY - Steady
STFR - Stratus fractus
STFRM - Stratiform
STG - Strong
STGLY - Strongly
STGR - Stronger
STGST - Strongest
STLT - Satellite
STM - Storm
STMS - Storms
STN - Station
STNS - Stations

STNRY - Stationary
SUB - Substitute
SUBTRPCL - Subtropical
SUF - Sufficient
SUFLY - Sufficiently
SUG - Suggest
SUGG - Suggesting
SUGS - Suggests
SUP - Supply
SUPG - Supplying
SUPR - Superior
SUPS - Supplies
SUPSD - Supersede
SUPSDG - Superseding
SUPSDS - Supersedes
SVG - Serving
SVR - Severe
SVRL - Several
SW - Southwest
SWD - Southward
SWWD - Southwestward
SW- - Light snow shower
SW+ - Heavy snow shower
SWLG - Swelling
SWLY - Southwesterly
SWOMCD - SELS Mesoscale Discussion
SWRN - Southwestern
SX - Stability index
SXN - Section
SXNS - Sections
SYNOP - Synoptic
SYNS - Synopsis
SYS - System
T - Thunder
TCNTL - Transcontinental
TCU - Towering cumulus
TDA - Today
TEMP - Temperature
THD - Thunderhead
THDR - Thunder

THK - Thick
THKNG - Thickening
THKNS - Thickness
THKR - Thicker
THKST - Thickest
THN - Thin
THNG - Thinning
THNR - Thinner
THNST - Thinnest
THR - Threshold
THRFTR - Thereafter
THRU - Through
THRUT - Throughout
THSD - Thousand
THTN - Threaten
THTND - Threatened
THTNG - Threatening
THTNS - Threatens
TIL - Until
TMPRY - Temporary
TMPRYLY - Temporarily
TMW - Tomorrow
TN - Tennessee
TNDCY - Tendency
TNDCYS - Tendencies
TNGT - Tonight
TNTV - Tentative
TNTVLY - Tentatively
TOPS - Tops
TOVC - Top of overcast
TPG - Topping
TRBL - Trouble
TRIB - Tributary
TRKG - Tracking
TRML -Terminal
TRMT - Terminate
TRMTD - Terminated
TRMTG - Terminating
TRMTS - Terminates
TRNSP - Transport
TRNSPG - Transporting

TROF - Trough
TROFS - Troughs
TROP - Tropopause
TRPCD - Tropical continental
TRPCL - Tropical
TRRN - Terrain
TRSN - Transition
TRW - Thunderstorm
TRW+ - Thunderstorm with heavy rain shower
TSFR - Transfer
TSFRD - Transferred
TSFRG - Transferring
TSFRS - Transfers
TSHWR - Thundershower
TSHWRS - Thundershowers
TSNT - Transient
TSQLS - Thundersqualls
TSTM - Thunderstorm
TSTMS - Thunderstorms
TS - Thunderstorm with snow
TS+ - Thunderstorm with heavy snow
TSW - Thunderstorm with snow showers
TSW+ - Thunderstorm with heavy snow showers
TURBC - Turbulence
TURBT - Turbulent
TWD - Toward
TWDS - Towards
TWI - Twilight
TWRG - Towering
TX - Texas
UDDF - Up and down drafts
UN - Unable
UNAVBL - Unavailable
UNEC - Unnecessary
UNKN - Unknown
UNL - Unlimited
UNRELBL - Unreliable
UNRSTD - Unrestricted

UNSATFY - Unsatisfactory
UNSBL - Unseasonable
UNSTBL - Unstable
UNSTDY - Unsteady
UNSTL - Unsettle
UNSTLD - Unsettled
UNUSBL - Unusable
UPDFTS - Updrafts
UPR - Upper
UPSLP - Upslope
UPSTRM - Upstream
URG - Urgent
USBL - Usable
UT - Utah
UVV - Upward vertical velocity
UVVS - Upward vertical velocities
UWNDS - Upper winds
VA - Virginia
VARN - Variation
VCNTY - Vicinity
VCOT - VFR conditions on top
VCTR - Vector
VDUC - VAS Data Utilization Center (NSSFC)
VFY - Verify
VFYD - Verified
VFYG - Verifying
VFYS - Verifies
VLCTY - Velocity
VLCTYS - Velocities
VLNT - Violent
VLNTLY - Violently
VLY - Valley
VLYS - Valleys
VMC - Visual meteorological conditions
VOL - Volume
VORT - Vorticity
VR - Veer
VRG - Veering
VRBL - Variable
VRISL - Vancouver Island, BC

VRS - Veers
VRT MOTN - Vertical Motion
VRY - Very
VSB - Visible
VSBY - Visibility
VSBYDR - Visibility decreasing rapidly
VSBYIR - Visibility increasing rapidly
VT - Vermont
VV - Vertical velocity
W - West
WA - Washington
WAA - Warm Air Advection
WBND - West bound
WDLY - Widely
WDSPRD - Widespread
WEA - Weather
WFO - Weather Forecast Office
WFOS - Weather Forecast Offices
WFP - Warm front passage
WI - Wisconsin
WIBIS - Will be issued
WINT - Winter
WK - Weak
WKDAY - Weekday
WKEND - Weekend
WKNG - Weakening
WKNS - Weakens
WKR - Weaker
WKST - Weakest
WKN - Weaken
WL - Will
WLY - Westerly
WND - Wind
WNDS - Winds
WNW - West-northwest
WNWLY - West-northwesterly
WNWRN - West-northwestern
WNWWD - West-northwestward
WO - Without
WPLTO - Western Plateau
WRM - Warm

WRMG - Warming
WRN - Western
WRMR - Warmer
WRMST - Warmest
WRMFNT - Warm front
WRMFNTL - Warm Frontal
WRNG - Warning
WRNGS - Warnings
WRS - Worse
WSHFT - Wind shift
WSHFTS - Wind Shifts
WSFO - Weather Service Forecast Office
WSFOS - Weather Service Forecast Offices
WSO - Weather Service Office
WSOS - Weather Service Offices
WSTCH - Wasatch Range
WSW - West-southwest
WSWLY - West-southwesterly
WSWRN - West-southwestern
WSWWD - West-southwestward
WTR - Water
WTRS - Waters
WTSPT - Waterspout
WTSPTS - Waterspouts
WUD - Would
WV - West Virginia
WVS - Waves
WW - Severe Weather Watch
WWAMKC - Status Report
WWD - Westward
WWS - Severe Weather Watches
WX - Weather
WY - Wyoming
XCP - Except
XPC - Expect
XPCD - Expected
XPCG - Expecting
XPCS - Expects
XPLOS - Explosive

XTND - Extend
XTNDD - Extended
XTNDG - Extending
XTRM - Extreme
XTRMLY - Extremely
YDA - Yesterday
YKN - Yukon
YLSTN - Yellowstone
ZL - Freezing drizzle
ZN - Zone
ZNS - Zones
ZR - Freezing rain

*"What freedom lies in flying, what Godlike power it gives
to men . . . I lose all consciousness in this strong unmortal
space crowded with beauty, pierced with danger.*
— *Charles A. Lindbergh (1902-1974)*

Aviation Acronyms & Abbreviations

AC - Advisory Circular
ACFT - Aircraft
AD - Airworthiness Directive
ADI - attitude direction indicator
ADF - automatic direction finder
ADIZ - air defense identification zone
A/FD - Airport/Facility Directory
AFM - Airplane Flight Manual
AFSS - Automated Flight Service Station
AGI - Advanced Ground Instructor
AGL - above ground level
AIM - Aeronautical Information Manual
AIRMET - Airman's Meteorological Info
ALS - approach light system
ALSF-1 - standard 2400' high-intensity approach lighting system with sequenced flashers (Category I configuration)
ALSF-2 - standard 2400' high-intensity approach lighting system with sequenced flashers (Category II configuration)
AME - aviation medical examiner
AMEL/S - airplane multi-engine land/sea
AOA - angle of attack
AOE - airport of entry
A&P - Airframe & Powerplant
APCH - approach
APU - auxiliary power unit
ARTC - air route traffic control
ARTCC - Air Route Traffic Control Center
ASL - above sea level (Canada)
ASOS - Automated Surface Observing System

ASR - airport surveillance radar
ASRS - Aviation Safety Reporting System
A/T - auto throttle
ATA - actual time of arrival
ATC - air traffic control
ATE - actual time enroute
ATIS - Automatic Terminal Info Service
ATP - Airline Transport Pilot
AWOS - Automatic Weather Observing System
BARO-VNAV - Barometric Vertical Navigation
BFR - Biennial Flight Review
BHP - Brake Horsepower
C - Celsius
CAS - calibrated airspeed
CAT II - Category II
CAVU - Ceiling and visibility unlimited
CDI - Course Deviation Indicator
CFI - Certificated Flight Instructor
CFIT - controlled flight into terrain
CFR - Code of Federal Regulations
CG - center of gravity
CHT - cylinder head temperature
CO - carbon monoxide
CO2 carbon dioxide
CONSOL or **CONSOLAN** - a low or medium frequency long range nav aid
CONUS - Continental U.S.
COP - change over point
CRM - crew resource management
CTAF - common traffic advisory frequency
CVR - cockpit voice recorder
CWA - center weather advisory

DA - density altitude
DALR - dry adiabatic lapse rate
DF/Steer - direction finding/steering
DG - directional gyro, heading indicator
DH - decision height
DME - distance measuring equipment compatible with TACAN
DP - Departure Procedure
DOT - Department of Transportation
DUAT(S) - Direct User Access Terminal (System)
DVFR - defense VFR
EAS - equivalent airspeed
EFAS - enroute flight advisory service
EFIS - electronic flight instrument system
EFC - expect further clearance
EGPWS - Enhanced Ground Proximity Warning System
EGT - exhaust gas temperature
ELT - emergency locator transmitter
ETA - estimated time of arrival
ETE - estimated time enroute
F - Fahrenheit
FA - area forecast report
FAA - Federal Aviation Administration
FAF - final approach fix
FAR - Federal Aviation Regulation
FAWP - Final Approach Waypoint

FBO - fixed base operator
FCC - Federal Communications Commission
FD - winds aloft report
FDC - Flight Data Center (FAA)
FDE - Fault Detection and Exclusion
FDR - flight data recorder
FE - flight engineer

FL - flight level
FLIP - Flight Information Publication
FM - fan marker
FMS - flight management system
FO - first officer
FOD - foreign object debris
FPM - feet per minute
FSDO - Flight Standards District Office (FAA)
FSS - flight service station
FYI - For your information
G - gravitational force
GA - general aviation
GMT - Greenwich Mean Time
GP - glide path
GPH - gallons per hour
GPS - global positioning system
GPU - ground power unit
GPWS - ground proximity warning system
GS - glide slope
HAA - height above airport
HAT - height above threshold
HF - high frequency
HIRL - high-intensity runway light system
HIWAS - Hazardous In-Flight Weather Advisory Service
HP - horsepower
HSI - horizontal situation indicator
HUD - heads-up display
IAF - initial approach fix
IAP - instrument approach procedure
IAS - indicated airspeed
ICAO - International Civil Aviation Organization.
IFR - instrument flight rules

IGI - Instrument Ground Instructor

ILS - instrument landing system

IM - ILS inner marker

IMC - instrument meteorological conditions

INOP - inoperative

INS - inertial navigation system

INT - intersection

ISA - International Standard Atmosphere

IR - military training route - instrument

KIAS - Knots, Indicated Airspeed

KHZ - kilohertz

LAA - local airport advisory

LAAS - local area augmentation system (GPS)

LAHSO - land & hold short operations

LAT - latitude

LBS - pounds

LDA - localizer-type directional aid

LF - low frequency

LFR - low-frequency radio range

LLWS - low level wind shear

LLWAS - low level wind shear alert system

LLZ - localizer

LMM - compass locator at middle marker

LNAV - lateral navigation

LOC - localizer

LOM - compass locator at outer marker

LONG - longitude

LOP - line of position

LORAN - long range radio aid to navigation

M - mach number

MAA - maximum authorized IFR altitude.

MAHP - Missed Approach Holding Point

MALS - medium intensity approach light system.

MALSR - medium intensity approach light system with runway alignment indicator lights

MAP - missed approach point

MAWP - Missed Approach Waypoint

MB - magnetic bearing

MCA - minimum crossing altitude

MDA - minimum descent altitude

MEA - minimum en route IFR altitude

MFD - multi-function display

MEI - multi engine instructor

MEL - minimum equipment list (91.213)

METAR - meteorological aviation routine report

MH - magnetic heading

MHZ - megahertz

MIRL - medium intensity runway lights

MM - ILS middle marker

MOA - military operations area

MOCA - minimum obstruction clearance altitude

MORA – minimum off-route altitude

MP - manifold pressure

MPH - miles per hour

MRA - minimum reception altitude

MSA - minimum safe altitude

MSL - mean sea level

MTR - military training route (see IR, VR)

MVA - minimum vectoring altitude

NA - not authorized

NACO - National Aeronautical

Charting Office (FAA)

NAS - National Airspace System

NAVAID - navigational aid, NDB, VOR, etc

NDB(ADF) - nondirectional beacon (automatic direction finder)

NDH - no damage history

NFCT - non federal control tower

NM - nautical mile

NOAA - National Oceanic and Atmospheric Administration

NOPT - no procedure turn required or authorized

NORDO - no radio

NOTAM - notice to airman

NPRM - notice of proposed rulemaking

NTSB - National Transportation Safety Board

NWS - National Weather Service

OAT - outside air temperature

OBS - omni-bearing selector

OEI - one engine inoperative

OM - ILS outer marker

OROCA - off route obstruction clearance altitude

OTS - out of service

PAPI - Precision Approach Path Indicator

PAR - precision approach radar

PCL - pilot controlled lighting

PF - pilot flying

PFD - primary flight display

PIC - pilot in command

PIREP - Pilot weather report

PM - pilot monitoring

PNF - pilot not flying

POH - pilot's operating handbook

PRM - Precision Runway Monitoring Approach

PSI - pounds per square inch

PT - procedure turn

PTS - practical test standards

PV - Prevailing Visibility

RA - Resolution Advisory

RAIL - runway alignment indicator light system

RAIM - receiver autonomous integrity monitoring (GPS)

RB - relative bearing

RBN - radio beacon

RCLM - runway centerline marking

RCLS - runway centerline light system

RCO - remote communications outlet

REIL - runway end identification lights

RMK - remark

RMI - radio magnetic indicator

RNAV - Area Navigation

RNP - required navigational performance

RPM - rotations per minute

RR - low or medium frequency radio range station

RV - radar vector

RVR - runway visual range as measured in the touchdown zone area

RVSM - reduced vertical separation minimum

RVV - Runway Visibility Value

SALR - saturated adiabatic lapse rate

SALS - short approach light system

SALSF - SALS with sequenced flashing lights

SAR - search and rescue

SDF - simplified directional facility

SIC - second in command
SID - Standard Instrument Departure
SIGMET - significant meteorological information
SLP - sea level pressure
SM - statute mile
SMOH - since major overhaul
SODA - statement of demonstrated ability
SOP - Standard operating procedure
SPOH - since prop overhaul
SSALS - simplified short approach light system.
SSALSR - simplified short approach light system with runway alignment indicator lights
STAR - standard terminal arrival route
STOL - short take off & landing
SUA - special use airspace
SVFR - special VFR
TACAN - tactical air navigational aid, a UHF military nav aid
TA - Traffic Alert
TAF - terminal aerodrome forecast
TAS - true airspeed
TAWS - Terrain Awareness and Warning System (see EGPWS)
TBO - time between overhaul
TC - true course
TCAS - traffic alert & collision avoidance system
TCH - threshold crossing height
TDZE - touchdown zone elevation
TDZL - touchdown zone lights
TEMPO - Temporary
TERPS - Terminal Instrument Procedures

TFR - temporary flight restriction
TH - true heading
TPA - traffic pattern altitude
TRACON - terminal radar approach control
TRSA - terminal radar service area
TSO - technical standard order
TTSN - total time since new
TVOR - terminal VHF omnirange station
TX - transmit or transponder
UHF - ultra-high frequency
UTC - universal coordinated time
VASI - visual approach slope indicator
VDP - visual descent point
VFR - visual flight rules
VHF - very high frequency
VMC - Visual meteorological conditions
VNAV - vertical navigation
VOR - VHF omnirange station.
VORTAC - co-located VOR and TACAN
VR - military training route - VFR
VSI - vertical speed indicator
VV - vertical visibility
WA - airmet
WAAS - wide area augmentation system (GPS)
WAC - world aeronautical chart
WH - hurricane advisory
WPT - waypoint
WS - sigmet
WST - convective sigmet
WW - Severe Weather Watch
WX - Weather
Z - zulu time

V-Speeds

V-Speeds

VA - design maneuvering speed.

VB - design speed for maximum gust intensity.

VC - design cruising speed.

VD - design diving speed.

VDF/MDF - demonstrated flight diving speed.

VF - design flap speed.

VFC/MFC - maximum speed for stability characteristics.

VFE - maximum flap extended speed.

VH - maximum speed in level flight with maximum continuous power.

VLE - maximum landing gear extended speed.

VLO - maximum landing gear operating speed.

VLOF - lift-off speed.

VMC - minimum control speed with the critical engine inoperative.

VMO/MMO - maximum operating limit speed.

VMU - minimum unstick speed.

VNE - never-exceed speed.

VNO - maximum structural cruising speed.

VP – Hydroplaning speed. For takeoff (wheels spinning) this speed is the square root of the tire pressure times 9. On landing (wheels not spinning) this speed is the square root of the tire pressure times 7.7.

VR - rotation speed.

VS - the stalling speed or the minimum steady flight speed at which the airplane is controllable.

VS0 - the stalling speed or the minimum steady flight speed in the landing configuration.

VS1 - the stalling speed or the minimum steady flight speed obtained in a specific configuration.

VTOSS - takeoff safety speed for Category A rotorcraft.

VX - speed for best angle of climb.

VY - speed for best rate of climb.

V1 - takeoff decision speed (formerly denoted as critical engine failure speed).

V2 - takeoff safety speed.

V2 min - minimum takeoff safety speed.

"Flying has torn apart the relationship of space and time. It uses our old clock but with new yardsticks."
— Charles A. Lindbergh (1902-1974)

Miscellaneous Acronyms & Contractions

Colors:

A - Amber
Be - Beige
Bk - Black
B - Blue
Br - Brown
Gd - Gold
Gy - Gray
G - Green
O - Orange
P - Purple
R - Red
S - Silver
T - Tan
V - Violet
W - White
Y - Yellow

Definitions

"**Shall**" is used in an imperative sense;
"**May**" is used in a permissive sense to
state authority or permission to do
the act prescribed, and the
words "no person may…" or "a
person may not…*" mean that no
person is required, authorized, or
permitted to do the act prescribed
"**Includes**" - "includes but is not limited
to"

"You're in charge but don't touch the controls."
— Shannon Lucid, recounting what
the two Russian cosmonauts told
her every time they left the Mir space
station for a space walk, 1996.

Aircraft Types

ICAO Aircraft Types

This page only includes "Light" category aircraft, i.e. a Boeing 777 won't be found here. This list is sorted by Manufacturer.

Type	Manufacturer & Model
A500	Adam A-500
A700	Adam A-700 AdamJet
AM3	Aermacchi / Macchi AM-3
F260	Aermacchi / Macchi SF-260
L90	Aermacchi / Macchi M-290TP Redigo
LA60	Aermacchi / Macchi AL-60
M308	Aermacchi / Macchi MB-308
M326	Aermacchi / Macchi MB-326
M326	Aermacchi / Macchi MB-326
M339	Aermacchi / Macchi MB-339
A270	AERO Ae-270 Ibis
AE45	AERO Ae-45 / Ae-145
L29	AERO L-29 Delfin
L39	AERO L-139 Albatross
L39	AERO L-39 Albatross
L59	AERO L-59
L60	AERO L-60 Brigadyr
MG15	AERO S-102
MG15	AERO S-103
AB11	Aero Boero AB-115
AB15	Aero Boero AB-150
AB18	Aero Boero AB-180
AB95	Aero Boero AB-95
A9	Aero Commander A-9 Ag Commander
AC50	Aero Commander 500 Commander 500
AC56	Aero Commander 560 Commander 560
AC68	Aero Commander 680F Commander 680F
AC6L	Aero Commander 680FL Grand Commander
AC72	Aero Commander Alti Cruiser
AC72	Aero Commander 720 Alti Cruiser
AC80	Aero Commander 680T / 680V Turbo Commander
CLB1	Aero Commander Ag Commander (B-1)
M200	Aero Commander Commander 200
SS2P	Aero Commander S-2 Ag Commander
VO10	Aero Commander 100 Commander 100
AR11	Aeronca 11 Chief
AR15	Aeronca 15 Sedan
CH7A	Aeronca 7 Champion
ALO2	Aerospatiale / SNIAS SA-318 Alouette 2

Type	Manufacturer & Model
ALO3	Aerospatiale / SNIAS SA-316 / SA-319 Alouette 3
AS50	Aerospatiale / SNIAS AS-350 SuperStar
AS50	Aerospatiale / SNIAS AS-550 Fennec
AS50	Aerospatiale / SNIAS AS-350 AStar
AS50	Aerospatiale / SNIAS AS-350 Ecureuil
AS55	Aerospatiale / SNIAS AS-355 Ecureuil 2
AS55	Aerospatiale / SNIAS AS-555 Fennec
AS55	Aerospatiale / SNIAS AS-355 TwinStar
AS65	Aerospatiale / SNIAS AS-565 Panther
AS65	Aerospatiale / SNIAS AS-365 Dauphin 2 / AS-366 Dolphin
FOUG	Aerospatiale / SNIAS CM-170R Magister
GAZL	Aerospatiale / SNIAS SA-341 / SA-342 Gazelle
LAMA	Aerospatiale / SNIAS SA-315 Lama
PUMA	Aerospatiale / SNIAS SA-330 Puma
S360	Aerospatiale / SNIAS SA-360 Dauphin
S601	Aerospatiale / SNIAS SN-601 Corvette
S65C	Aerospatiale / SNIAS SA-365C Dauphin 2
TB30	Aerospatiale / SNIAS TB-30 Epsilon
YK52	Aerostar Yak-52
PTS1	Aerotek Pitts S-1 Special
PTS2	Aerotek Pitts S-2 Special
CT4	AESL CT-4 Airtrainer
TOUR	AESL Airtourer
AA5	AGAC AG-5 Tiger
G64T	Ag-Cat G-164 Super Turbine
A109	Agusta / AgustaWestland A-109
A119	Agusta / AgustaWestland A-119 Koala
A129	Agusta / AgustaWestland A-129 Mangusta
A139	Agusta / AgustaWestland AB139 / AW139
B206	Agusta / AgustaWestland AB-206 JetRanger / LongRanger
B212	Agusta / AgustaWestland AB-212
B412	Agusta / AgustaWestland AB-412 Griffon
B47G	Agusta / AgustaWestland AB-47G
B47J	Agusta / AgustaWestland AB-47J
CH1	AIDC A-CH-1 Chung-Tsing
ERCO	Air Products Aircoupe
AT3P	Air Tractor AT-300 / AT-301 / AT-401
AT3T	Air Tractor AT-302 / AT-400 / AT-402
AT5P	Air Tractor AT-501
AT5T	Air Tractor AT-502 / AT-503
AT6T	Air Tractor AT-602
R200	Alpha R-2160 Alpha 160
ROND	Ambrosini F-4 Rondone
ROND	Ambrosini Rondone
AA1	American AA-1

Type	Manufacturer & Model
AA5	American AA-5
AA5	American AA-5
AC90	American 690 Jetprop Commander 840 / 900
AC95	American 695 Jetprop Commander 1000
AC95	American Jetprop Commander 980 / 1000
G164	American G-164 Ag-Cat
G64T	American G-164 Turbo Ag-Cat
GA7	American GA-7 Cougar
BL8	American Champion 8 Super Decathlon
BL8	American Champion 8 Scout
CH7A	American Champion 7ECA Citabria Aurora
CH7A	American Champion 7ACA Champ
CH7B	American Champion Citabria Adventure
CH7B	American Champion Citabria Explorer
AN2	Antonow / Antonov An-2
AN28	Antonow / Antonov An-28
AN3	Antonow / Antonov An-3
AR79	Arado Ar-79
AM3	Atlas AM-3 Bosbok
LA60	Atlas AL-60 Kudu
M326	Atlas MB-326 Impala
PC7	Atlas PC-7 Astra
ACAR	Auster J-5 Autocar
ADVE	Auster J-5 Adventurer
AIGT	Auster J-5 Aiglet Trainer
ALPI	Auster J-5 Alpine
AUJ2	Auster J-2 Arrow
AUJ4	Auster J-4 Archer
AUS5	Auster Auster 5 Alpha
AUS6	Auster Auster AOP6
AUS7	Auster Auster T7
AUS9	Auster Auster AOP9
D4	Auster D-4
D5	Auster D-5
D6	Auster D-6
J1	Auster J-1 Aiglet
F8L	Aviamilano F-8L Falco
NIBB	Aviamilano F-14 Nibbio
P19	Aviamilano P-19 Scricciolo
HUSK	Aviat A-1 Husky
M110	Aviat 110 Specal
PTS1	Aviat Pitts S-1 Special
PTS2	Aviat Pitts S-2 Special
PTSS	Aviat S-1-11 Super Stinker
NIPR	Avions Fairey T-66 Nipper

Type	Manufacturer & Model	
TIJU	Avions Fairey	Tipsy Junior
TIPB	Avions Fairey	Tipsy BC
TIPB	Avions Fairey	Tipsy Belfair
TIPB	Avions Fairey	Tipsy Trainer
TIPB	Avions Fairey	Tipsy B
A504	AVRO	504
ANSN	AVRO	652 Anson
AVIN	AVRO	594 / 616 Avian
TUTR	AVRO	621 Tutor
A660	Ayres	S-2R-T660 Turbo Thrush
SS2P	Ayres	Thrush
SS2P	Ayres	Bull Thrush
SS2T	Ayres	Turbo Thrush (S-2R-G/T except T660)
JPRO	BAC	167 Strikemaster
JPRO	BAC	145 Jet Provost
BDOG	BAe	Bulldog
JPRO	BAe	BAC-167 Strikemaster
AIRD	Beagle	A-109 Airedale
AU11	Beagle	E-3 Auster AOP11
AUS6	Beagle	A-61 Terrier / Tugmaster
BASS	Beagle	B-206
BDOG	Beagle	B-125 Bulldog
D4	Beagle	D-4
D5	Beagle	D-5
D5	Beagle	D-5
D5	Beagle	D-5 Husky
D6	Beagle	D-6
PUP	Beagle	B-121 Pup
B18T	Beech	18 (turbine)
B350	Beech	300 (B300) Super King Air 350
B36T	Beech	36 Bonanza (turbine)
BE10	Beech	100 King Air
BE17	Beech	17 Staggerwing
BE18	Beech	18
BE20	Beech	200 Super King Air
BE20	Beech	1300 Commuter
BE23	Beech	23 Musketeer
BE23	Beech	23 Sundowner
BE24	Beech	24 Sierra
BE24	Beech	24 Musketeer Super
BE30	Beech	300 Super King Air
BE33	Beech	33 Debonair
BE33	Beech	33 Bonanza
BE35	Beech	35 Bonanza
BE36	Beech	36 Bonanza

Type	Manufacturer & Model	
BE50	Beech	50 Twin Bonanza
BE50	Beech	Seminole (U-8D/E/G)
BE55	Beech	55 Baron
BE58	Beech	58 Baron
BE60	Beech	60 Duke
BE65	Beech	65 Queen Air
BE70	Beech	70 Queen Air
BE76	Beech	76 Duchess
BE77	Beech	77 Skipper
BE80	Beech	80 Queen Air
BE88	Beech	88 Queen Air
BE95	Beech	95 Travel Air
BE99	Beech	99 Airliner
BE9L	Beech	90 King Air
BE9T	Beech	90 (F90) King Air
STAR	Beech	2000 Starship
T34P	Beech	45 Mentor
T34P	Beech	T-34 Mentor
T34T	Beech	T-34C Turbo Mentor
U21	Beech	U-21A Ute
P39	Bell Aircraft	P-39 Airacobra
P63	Bell Aircraft	P-63 Kingcobra
B06	Bell Helicopter	JetRanger / OH-58 Kiowa
B06T	Bell Helicopter	206LT TwinRanger
B212	Bell Helicopter	UH-1N
B212	Bell Helicopter	212 Twin Two-Twelve
B222	Bell Helicopter	222
B230	Bell Helicopter	230
B407	Bell Helicopter	407
B412	Bell Helicopter	CH-146 Griffon
B412	Bell Helicopter	412
B427	Bell Helicopter	427
B429	Bell Helicopter	429 GlobalRanger
B430	Bell Helicopter	430
B47G	Bell Helicopter	47
B47J	Bell Helicopter	47J Ranger
HUCO	Bell Helicopter	209 HueyCobra
UH1	Bell Helicopter	UH-1 Iroquois
UH1	Bell Helicopter	205
UH1	Bell Helicopter	210
B14A	Bellanca	14 Cruisair / Cruisemaster / Junior
B14C	Bellanca	14 Bellanca 260A / B / C
BL17	Bellanca	17 Super Viking
BL19	Bellanca	19 Skyrocket
BL8	Bellanca	8 Decathlon

Type	Manufacturer & Model	
BL8	Bellanca	8 Scout
CH7A	Bellanca	Champ
CH7A	Bellanca	Citabria (7ECA)
CH7B	Bellanca	Citabria (7GCBC/7KCAB)
T250	Bellanca	T-250 Aries
F156	Benes-Mraz	K-65 Cap
SOKL	Benes-Mraz	M-1 Sokol
B103	Beriev	Be-103 Bekas
SA20	Beriev	SA-20
SWOR	Blackburn	Swordfish
BL11	Bleriot	11
ST75	Boeing	PT-18 Kaydet
ST75	Boeing	PT-13 Kaydet
ST75	Boeing	PT-17 Kaydet
ST75	Boeing	75 Kaydet
ST75	Boeing	PT-27 Kaydet
H66	Boeing-Sikorsky	RAH-66 Comanche
B209	Bölkow	BO-209 Monsun
JUNR	Bölkow	BO-208 Junior
KL07	Bölkow	Kl-107
BRB2	Brantly	B-2
BR14	Breguet	14 Replica
BFIT	Bristol	Fighter
BFIT	Bristol	F-2B Fighter
BLEN	Bristol	149 Blenheim
BLEN	Bristol	149 Bolingbroke
SYCA	Bristol	171 Sycamore
SYCA	Bristol	Sycamore
BN2P	Britten Norman	BN-2 Islander
BN2T	Britten Norman	BN-2T Defender 4000 / Turbine Islander
TRIS	Britten Norman	BN-2A Mk3 Trislander
BU31	Bücker	Bü-131 Jungmann
BU33	Bücker	Bü-133 Jungmeister
BU81	Bücker	Bü-181 Bestmann
CA19	CAC	Boomerang
CA25	CAC	CA-25 Winjeel
M326	CAC	MB-326
M326	CAC	CA-30
P51	CAC	CA-18 Mustang
WIRR	CAC	CA-16 Wirraway
BL8	Campion	8 Decathlon
CH40	Campion	402 Lancer
CH7A	Campion	Traveler
CH7A	Campion	Citabria (7ECA)
CL41	Canadair	CT-114 Tutor

Type	Manufacturer & Model	
CL41	Canadair	CL-41 Tutor
BABY	Canadian Home Rotors	Baby Belle / Safari
CP10	CAP	CAP-10
CP22	CAP	CAP-222
BU31	CASA	1-131
BU33	CASA	1-133
C101	CASA	C-101 Aviojet
PILL	CASA	T-35 Pillán
PILL	CASA	E-26 Tamiz
S223	CASA	223 Flamingo
SATA	CASA	HA-220 Super Saeta
C270	Caudron	C-272 / C-275 Luciole
C270	Caudron	C-275 Luciole
CG3	Caudron	G-3
JN76	Caudron	CR-760 Cyclone Replica
JN76	Caudron	JN-760 Cyclone Replica
T6	CCF	T-6 Harvard
CE43	Cerva	CE-43 Guepard
A37	Cessna	318 / A-37 Dragonfly
C02T	Cessna	402 (turbine)
C04T	Cessna	404 (turbine)
C06T	Cessna	206 (turbine)
C07T	Cessna	207 (turbine)
C10T	Cessna	P210 (turbine)
C120	Cessna	120
C140	Cessna	140
C14T	Cessna	414 (turbine)
C150	Cessna	A150 Aerobat
C150	Cessna	150
C152	Cessna	A152 Aerobat
C152	Cessna	152
C162	Cessna	162 SkyCatcher
C170	Cessna	170
C172	Cessna	172
C175	Cessna	175
C177	Cessna	177
C180	Cessna	180
C182	Cessna	182
C182	Cessna	T182 Turbo Skylane
C185	Cessna	185 Skywagon
C185	Cessna	A185 AgCarryall
C188	Cessna	188 AgPickup / AgWagon
C190	Cessna	190
C195	Cessna	195
C205	Cessna	205

Type	Manufacturer & Model
C206	Cessna T206 Turbo Stationair
C206	Cessna 206 Stationair / Super Skywagon
C207	Cessna T207 Turbo Skywagon 207
C207	Cessna T207 Turbo Stationair 7 / 8
C207	Cessna 207 Stationair 7 / 8
C207	Cessna 207 Skywagon 207
C208	Cessna 208 Caravan 1
C210	Cessna 210 Centurion
C21T	Cessna 421 (turbine)
C25A	Cessna 525A Citation CJ2
C25B	Cessna 525B Citation CJ3
C303	Cessna T303 Crusader
C310	Cessna 310
C320	Cessna 320 Executive Skyknight
C335	Cessna 335
C336	Cessna 336 Skymaster
C337	Cessna 337 Super Skymaster
C340	Cessna 340
C402	Cessna 401
C402	Cessna 402
C404	Cessna 404 Titan
C411	Cessna 411
C414	Cessna 414
C421	Cessna 421 Golden Eagle
C425	Cessna 425 Corsair
C425	Cessna 425 Conquest 1
C441	Cessna 441 Conquest 2
C441	Cessna Conquest
C500	Cessna 500 Citation 1
C501	Cessna 501 Citation 1SP
C510	Cessna 510 Citation Mustang
C525	Cessna 525 Citation CJ1
C526	Cessna 526 CitationJet
C550	Cessna 550 Citation Bravo
C550	Cessna S550 Citation S2
C550	Cessna 550 Citation 2
C551	Cessna 551 Citation 2SP
C72R	Cessna 172RG Cutlass RG
C77R	Cessna 177RG Cardinal RG
C82R	Cessna R182 Skylane RG
C82R	Cessna TR182 Turbo Skylane RG
CMAS	Cessna Airmaster
F406	Cessna F406 Caravan 2
O1	Cessna Bird Dog
P210	Cessna P210 Pressurized Centurion

Type	Manufacturer & Model
P337	Cessna P337 Pressurized Skymaster
T37	Cessna 318 / T-37
AS65	Chance-Vought SA-366 Panther 800
MG17	Chengdu JJ-5
MG17	Chengdu FT-5
EAGL	Christen Eagle
PTS1	Christen Pitts S-1 Special
PTS2	Christen Pitts S-2 Special
SR20	Cirrus SR-20
SR22	Cirrus SR-22
AC11	Commander Commander 114
AC11	Commander Commander 115
L13	Convair L-13
VALI	Convair BT-13 Valiant
P40	Curtiss P-40 Warhawk
P40	Curtiss Kittyhawk
FLAM	Dassault Flamant
DH60	De Havilland DH-60 Moth
DH80	De Havilland DH-80 Puss Moth
DH82	De Havilland DH-82 Tiger Moth
DH83	De Havilland DH-83 Fox Moth
DH85	De Havilland DH-85 Leopard Moth
DH87	De Havilland DH-87 Hornet Moth
DH88	De Havilland DH-88 Comet Replica
DH89	De Havilland DH-89 Dragon Rapide
DH90	De Havilland DH-90 Dragonfly
DH94	De Havilland DH-94 Moth Minor
DHC1	De Havilland DHC-1 Chipmunk
DOVE	De Havilland DH-104 Dove
HERN	De Havilland DH-114 Heron
VAMP	De Havilland DH-115 Vampire
VAMP	De Havilland DH-100 Vampire
DHA3	De Havilland Australia DHA-3 Drover
DH2T	De Havilland Canada DHC-2 Mk3 Turbo Beaver
DH3T	De Havilland Canada DHC-3 Turbo Otter
DHC2	De Havilland Canada DHC-2 Mk1 Beaver
DHC3	De Havilland Canada DHC-3 Otter
DHC6	De Havilland Canada DHC-6 Twin Otter
PC7	Denel PC-7 Astra
DWD2	Dewoitine D-26
DWD2	Dewoitine D-27
DA40	Diamond DA-40
DA42	Diamond DA-42
DA50	Diamond DA-50
DJET	Diamond D-Jet

Type	Manufacturer & Model
DV20	Diamond Katana (DA-20/DV-20)
D228	Dornier 228
D28D	Dornier Do-28D Skyservant
D28T	Dornier 128-6 Turbo Skyservant
DO27	Dornier Do-27
DO28	Dornier Do-28A / Do-28B
SBD	Douglas SBD Dauntless
EA50	Eclipse Eclipse 500
OPCA	Edgley / Optica EA-7 Optica
C365	EKW / K+W C-3605
DWD2	EKW / K+W D-27
DWD2	EKW / K+W D-26
E110	EMBRAER EMB-110 Bandeirante
E110	EMBRAER EMB-111 Bandeirulha
E121	EMBRAER EMB-121 Xingu
E314	EMBRAER EMB-314 Super Tucano
E50P	EMBRAER EMB-500 Phenom 100
E55P	EMBRAER EMB-505 Phenom 300
IPAN	EMBRAER EMB-200 Ipanema
M326	EMBRAER EMB-326 Xavante
M326	EMBRAER AT-26 Xavante
P28A	EMBRAER Carioquinha
P28A	EMBRAER EMB-712
P28A	EMBRAER Tupi
P28A	EMBRAER EMB-712 Tupi
P28B	EMBRAER EMB-710 Carioca
P28R	EMBRAER EMB-711B Corisco
P28T	EMBRAER EMB-711ST Corisco 2 Turbo
P32R	EMBRAER EMB-721 Sertanejo
PA25	EMBRAER PA-25 Pawnee
PA31	EMBRAER EMB-820 Navajo
PA32	EMBRAER EMB-720 Minuano
PA34	EMBRAER EMB-810 Seneca
TUCA	EMBRAER T-27 Tucano
TUCA	EMBRAER EMB-312 Tucano
EN28	Enstrom F-28
EN28	Enstrom 280 Shark
EN48	Enstrom 480
ERCO	ERCO Ercoupe
AS50	Eurocopter AS-350 Ecureuil
AS50	Eurocopter AS-350 AStar
AS50	Eurocopter AS-550 Fennec
AS50	Eurocopter AS-350 SuperStar
AS55	Eurocopter AS-355 TwinStar
AS55	Eurocopter AS-355 Ecureuil 2

Type	Manufacturer & Model
AS55	Eurocopter AS-555 Fennec
AS65	Eurocopter AS-565 Panther
AS65	Eurocopter AS-365 Dauphin 2 / AS-366 Dolphin
BK17	Eurocopter BK-117
EC20	Eurocopter EC-120 Colibri
EC30	Eurocopter EC-130
EC35	Eurocopter EC-635
EC35	Eurocopter EC-135
EC45	Eurocopter EC-145
EC45	Eurocopter BK-117C-2
EC55	Eurocopter EC-155
TIGR	Eurocopter EC-665 Tiger / Tigre
BK17	Eurocopter-Kawasaki BK-117
EC45	Eurocopter-Kawasaki BK-117C-2
EC45	Eurocopter-Kawasaki EC-145
VP2	Evans VP-2 Volksplane
E200	EXTRA EA-200
E230	EXTRA EA-230
E300	EXTRA EA-300
E400	EXTRA EA-400
E500	EXTRA EA-500
FA11	Fairchild F-11 Husky
FA24	Fairchild UC-61 Argus
FA24	Fairchild UC-61 Forwarder
FA62	Fairchild Cornell
KR34	Fairchild KR-34 Challenger
KR34	Fairchild Challenger
SW3	Fairchild Merlin 3
FH11	Fairchild-Hiller FH-1100
UH12	Fairchild-Hiller OH-23 Raven
SW3	Fairchild-Swearingen Merlin 3
FFLY	Fairey Firefly
SWOR	Fairey Swordfish
AS02	FFA AS-202 Bravo
AS2T	FFA AS-202-32TP Turbine Bravo
G46	Fiat G-46
G59	Fiat G-59
F156	Fieseler Fi-156 Storch
FU24	Fletcher FU-24
C172	FMA P-172
C182	FMA A-182
C182	FMA A-182
C188	FMA A-A188 AgTruck
IA46	FMA IA-46 Ranquel
IA46	FMA IA-46 Super Ranquel

Type	Manufacturer & Model	
IA51	FMA	IA-51 Tehuelche
IA51	FMA	IA-51 Tehuelche
IA58	FMA	IA-58 Pucará
IA63	FMA	IA-63 Pampa
FW44	Focke-Wulf	Fw-44 Stieglitz
FW90	Focke-Wulf	Fw-190 Replica
P149	Focke-Wulf	FWP-149
D7	Fokker	D-7 Replica
D8	Fokker	D-8 Replica
DR1	Fokker	Dr-1 Replica
S11	Fokker	S-11 Instructor
GNAT	Folland	Fo-144 Gnat
TRIM	Ford	5-AT Tri-Motor
TRIM	Ford	4-AT Tri-Motor
ERCO	Forney / Fornaire Aircoupe	
FOUG	Fouga	CM-170R Magister
RF10	Fournier RF-10	
RF47	Fournier RF-47	
RF6	Fournier RF-6	
RF9	Fournier RF-9	
KM2	Fuji	TL-1
KM2	Fuji	KM-2
O1	Fuji	Bird Dog
RC70	Fuji	FA-300
SUBA	Fuji	FA-200 Aero Subaru
T1	Fuji	T-1
T34P	Fuji	T-3 Mentor
T34T	Fuji	T-3Kai
T5	Fuji	KM-2Kai
T5	Fuji	T-5
T7	Fuji	T-7
NOMA	GAF	Nomad
JP10	GAP	JP100 Kestrel
ARON	General Avia	F-220 Airone
PEGA	General Avia	F-20 Pegaso
PICO	General Avia	F-15 Delfino
PICO	General Avia	F-15 Picchio
PINO	General Avia	F-22 Pinguino
GA20	Gippsland	GA-200 Fatman
GA20	Gippsland	Fatman
GA8	Gippsland	GA-8 Airvan
GC1	Globe	GC-1 Swift
GAUN	Gloster	Gauntlet
GLAD	Gloster	Gladiator
G109	Grob	G-109

Type	Manufacturer & Model	
G115	Grob	G-115
G120	Grob	G-120
G140	Grob	G-140TP
G15T	Grob	G-115T Acro
G160	Grob	G-160 Ranger
GF20	Grob	GF-200
GSPN	Grob	G-180 SPn Utility Jet
BCAT	Grumman	F8F Bearcat
F3F	Grumman	G-32 Replica
F3F	Grumman	G-11 Replica
F3F	Grumman	F3F Replica
F9F	Grumman	F9F Panther
G164	Grumman	G-164 Ag-Cat
G21	Grumman	G-21A Goose
G44	Grumman	G-44 Widgeon
G64T	Grumman	G-164 Turbo Ag-Cat
G73	Grumman	G-73 Mallard
G73T	Grumman	G-73T Turbo Mallard
HCAT	Grumman	F6F Hellcat
WCAT	Grumman	F4F Wildcat
ALH	HAL	Dhruv
ALO3	HAL	SA-316 / SE-3160 Chetak / Chetan
AR11	HAL	HUL-26 Pushpak
D228	HAL	228
HA31	HAL	HA-31 Basant
HT16	HAL	HJT-16 Kiran
HT2	HAL	HT-2
HT32	HAL	HPT-32 Deepak
HT36	HAL	HJT-36 Sitara
LAMA	HAL	SA-315 Cheetah / Cheetal / Lancer
AS65	HAMC / Harbin	Z-9 Haitun
EC20	HAMC / Harbin	HC-120
MI4	HAMC / Harbin	Z-5
Y11	HAMC / Harbin	Y-11
Y12	HAMC / Harbin	Y-12
JS1	Handley Page	HP-137 Jetstream 1
JS20	Handley Page	Jetstream 200
JS20	Handley Page	HP-137 Jetstream 200
CYGT	Hawker	Cygnet
FURY	Hawker	Fury
FURY	Hawker	Sea Fury
HIND	Hawker	Hind
HURI	Hawker	Hurricane
HURI	Hawker	Sea Hurricane
PRM1	Hawker-Beechcraft	390 Premier 1

Aviation Acronyms, Contractions, & Mnemonics (PocketLearning)

Type	Manufacturer & Model
TEX2	Hawker-Beechcraft CT-156 Harvard 2
TEX2	Hawker-Beechcraft T-6 Texan 2
DOVE	Hawker-Siddeley DH-104 Dove
GNAT	Hawker-Siddeley Gnat
HERN	Hawker-Siddeley DH-114 Heron
COUR	Helio Courier
STLN	Helio Stallion
TCOU	Helio Twin Courier
BU81	Heliopolis Gomhouria
BE33	HESA Parastu
PC7	HESA S-68
S278	HESA Shahed 278
H12T	Hiller UH-12ET
UH12	Hiller OH-23 Raven
HI27	Hirth Hi-27 Acrostar
ME09	Hispano HA-1112 Buchon
S223	Hispano 223 Flamingo
SATA	Hispano HA-200 Saeta
SATA	Hispano HA-220 Super Saeta
HDJT	Honda HA-420 HondaJet
H269	Hughes 300
H269	Hughes TH-55 Osage
H269	Hughes 269
H500	Hughes 369
H500	Hughes OH-6 Cayuse
H500	Hughes 500
JPRO	Hunting P-84 Jet Provost
PEMB	Hunting P-66 Pembroke
PPRO	Hunting P-56 Provost
JPRO	Hunting Percival P-84 Jet Provost
PEMB	Hunting Percival P-66 Pembroke
PPRO	Hunting Percival P-56 Provost
ARVA	IAI Arava
IR46	IAR IAR-46
IS28	IAR IS-28M2
PUMA	IAR IAR-330 Puma
A270	Ibis Ae-270 Ibis
I103	Ilyuschin Il-103
CH2T	JAI CH-2000 Sama
CH80	JAI CH-8000 Hawk 1
D11	Jodel D-11
D11	Jodel D-120
D140	Jodel D-140 Abeille
D140	Jodel D-140 Mousquetaire
D150	Jodel D-150 Mascaret

Type	Manufacturer & Model	
D18	Jodel	D-18
B427	KAI	SB-427
KT1	KAI	KT-1 Woong-Bee
KT1	KAI	KO-1 Woong-Bee
H2	Kaman	SH-2 Seasprite
H43A	Kaman	HH-43A
H43B	Kaman	HH-43B Huskie
KMAX	Kaman	K-1200 K-Max
K126	Kamov	Ka-126
K226	Kamov	Ka-226
KA26	Kamov	Ka-26
KA62	Kamov	Ka-62
KA62	Kamov	Ka-60
B47G	Kawasaki	47G
BK17	Kawasaki	BK-117
EC45	Kawasaki	BK-117C-2
KH4	Kawasaki	KH-4
OH1	Kawasaki	OH-1
A2RT	Kazan	Ansat 2RT
ANST	Kazan	Ansat
KL07	Klemm	Kl-107
KL25	Klemm	L-25
KL35	Klemm	Kl-35
LA25	Lake Aircraft	LA-250 / LA-270
LA4	Lake Aircraft	LA-4 / LA-200 Buccaneer
LJ23	Learjet	23
LJ24	Learjet	24
LJ25	Learjet	25
AE45	LET	Super Aero 145
AE45	LET	Aero 45
L200	LET	L-200 Morava
L410	LET	L-420 Turbolet
L410	LET	L-410 Turbolet
YK11	LET	C-11
Z37P	LET	Z-237 Cmelák
Z37P	LET	Z-37 Cmelák
L10	Lockheed	L-10 Electra
L12	Lockheed	L-12 Electra Junior
LA60	Lockheed	LASA-60
LA60	Lockheed-Azcarate	LASA-60 Santa Maria
L11	Luscombe	11A Sedan
L8	Luscombe	50
L8	Luscombe	8
SP18	Luscombe	11E Spartan 185
BROU	Max Holste	MH-1521 Broussard

Type	Manufacturer & Model	
B105	MBB	BO-105
BK17	MBB	BK-117
S223	MBB	223 Flamingo
S223	MBB	223 Flamingo
BK17	MBB-Kawasaki	BK-117
MD52	McDonnell Douglas	MH-6J
MD52	McDonnell Douglas	AH-6J
MD52	McDonnell Douglas	MD-520N
MD52	McDonnell Douglas	MD-530N
MD60	McDonnell Douglas	MD-600N
EXPL	MD Helicopters	MD-902 Explorer
H500	MD Helicopters	MD-500
H500	MD Helicopters	MD-530
MD52	MD Helicopters	MD-520N
MD60	MD Helicopters	MD-600N
ME08	Messerschmitt	Bf-108 Taifun
ME09	Messerschmitt	Bf-109
ME62	Messerschmitt	Me-262 Replica
JUNR	MFI	MFI-9 Junior
MG15	Mikoyan MiG-15	
MG17	Mikoyan MiG-17	
MGAT	Mikoyan MiG-AT	
MI2	Mil	Mi-2
MI34	Mil	Mi-34
MI4	Mil	Mi-4
FALM	Miles	M-3 Falcon Major
GEMI	Miles	M-65 Gemini
M2HK	Miles	M-2 Hawk Speed Six
M2HK	Miles	M-2 Hawk Major
MAGI	Miles	M-14 Magister
MESS	Miles	M-38 Messenger
MONA	Miles	M-17 Monarch
WHIT	Miles	M-11 Whitney Straight
MH20	Mitsubishi	MH-2000
MU2	Mitsubishi	MU-2
RP1	Mitsubishi	RP-1
ZERO	Mitsubishi	A6M Zero
ERCO	Mooney Aircoupe	
M10	Mooney M-10 Cadet	
M20P	Mooney M-20	
M20T	Mooney M-20K 252TSE / M-20M	
M22	Mooney M-22 Mustang	
MITE	Mooney M-18	
F156	Morane-Saulnier	Criquet
MS23	Morane-Saulnier	MS-230

Type	Manufacturer & Model	
MS31	Morane-Saulnier	MS-317
MS31	Morane-Saulnier	MS-315
MS73	Morane-Saulnier	MS-733 Alcyon
MS76	Morane-Saulnier	MS-760 Paris
MSAI	Morane-Saulnier	MoS-29
MSAI	Morane-Saulnier	MoS-27
MSAI	Morane-Saulnier	MoS-30
MSAI	Morane-Saulnier	AI
RALL	Morane-Saulnier	Super Rallye
RALL	Morane-Saulnier	Rallye Commodore
RALL	Morane-Saulnier	Rallye Club
RALL	Morane-Saulnier	Rallye 105
CP10	Mudry	CAP-10
CP20	Mudry	CAP-20
CP21	Mudry	CAP-21
CP23	Mudry	CAP-230
D140	Mudry	D-140 Mousquetaire
M101	Myasishchev	M-101
M203	Myasishchev	M-203 Barsuk
AN2	Nanchang	Y-5
CJ6	Nanchang	CJ-6
K8	Nanchang	JL-8 Karakorum
K8	Nanchang	K-8 Karakorum
N5	Nanchang	N-5
YK18	Nanchang	CJ-5
NI28	Nieuport	28 Replica
NORS	Noorduyn	Norseman
T6	Noorduyn	AT-16 Harvard
ME08	Nord / SNCAN	1001 Pingouin 1
ME08	Nord / SNCAN	1002 Pingouin 2
N110	Nord / SNCAN	1100 Noralpha
N120	Nord / SNCAN	1200 Norecrin
N320	Nord / SNCAN	3202
N340	Nord / SNCAN	3400
NC85	Nord / SNCAN	NC-858
NC85	Nord / SNCAN	NC-854
SV4	Nord / SNCAN	SV-4
NAVI	North American	Navion
P51	North American	P-51 Mustang
T2	North American	T-2 Buckeye
T28	North American	T-28 Trojan
T6	North American	T-6 Texan
T6	North American	SNJ Texan
T6	North American	AT-6 Texan
V10	North American	OV-10 Bronco

Aviation Acronyms, Contractions, & Mnemonics (PocketLearning)

Type	Manufacturer & Model
YALE	North American Yale
A9	North American Rockwell A-9 Quail Commander
AC11	North American Rockwell 112 Commander 112
AC50	North American Rockwell 500 Commander 500
AC50	North American Rockwell 500 Shrike Commander
AC6L	North American Rockwell Grand Commander
AC6L	North American Rockwell 685 Commander 685
AC6L	North American Rockwell 680FL Courser Commander
AC80	North American Rockwell 680V / 680W / 681 Turbo Commander
AC90	North American Rockwell 690 Turbo Commander 690
CLB1	North American Rockwell B-1 Snipe Commander
LARK	North American Rockwell 100 Lark Commander
M200	North American Rockwell 200 Commander 200
SS2P	North American Rockwell Ag Commander
SS2P	North American Rockwell Thrush Commander
T2	North American Rockwell T-2 Buckeye
V10	North American Rockwell OV-10 Bronco
VO10	North American Rockwell Darter Commander
T38	Northrop T-38 Talon
L40	Orlican L-40 Meta Sokol
CRES	PAC Cresco
CT4	PAC CT-4 Airtrainer
P750	PAC 750XL
OSCR	Partenavia P-64 Oscar
OSCR	Partenavia P-66 Charlie
P57	Partenavia P-57 Fachiro 2
P68	Partenavia P-68 Observer
P68	Partenavia P-68 Victor
P68T	Partenavia AP-68TP-300 Spartacus
VTOR	Partenavia AP-68TP-600 Viator
PEMB	Percival P-66 Pembroke
PPRO	Percival P-56 Provost
PRCE	Percival P-57 Sea Prince
PREN	Percival P-40 Prentice
PROC	Percival Proctor
VGUL	Percival Vega Gull
P136	Piaggio P-136
P148	Piaggio P-148
P149	Piaggio P-149
P180	Piaggio P-180 Avanti
P66P	Piaggio P-166
P66T	Piaggio P-166DL3
P66T	Piaggio P-166DP1
CP13	Piel Super Emeraude (CP-1310/1315/1330)
CP30	Piel Emeraude

Type	Manufacturer & Model	
CP32	Piel	Super Emeraude (CP-320/325/328)
CP60	Piel	Diamant
CP75	Piel	Béryl
CP80	Piel	Zéphir
CP90	Piel	Pinocchio
SAPH	Piel	CP-1320 Saphir
PC12	Pilatus	PC-12
PC21	Pilatus	PC-21
PC6P	Pilatus	PC-6 Porter
PC6T	Pilatus	PC-6A/B/C Turbo-Porter
PC7	Pilatus	PC-7 Turbo Trainer
PC9	Pilatus	PC-9
PP2	Pilatus	P-2
PP3	Pilatus	P-3
AEST	Piper	PA-60 Aerostar
J2	Piper	J-2 Cub
J2	Piper	Cub (J-2)
J3	Piper	Cub (J-3/L-4/NE)
J4	Piper	J-4 Cub Coupe
J5	Piper	J-5 Cub Cruiser
J5	Piper	L-14 Cub Cruiser
P28A	Piper	Cherokee (PA-28-140/150/160/180)
P28B	Piper	Cherokee (PA-28-235)
P28R	Piper	PA-28R-200 Cherokee Arrow
P28R	Piper	PA-28R-180 Cherokee Arrow
P28R	Piper	PA-28R-201T Turbo Cherokee Arrow 3
P28R	Piper	PA-28R-201 Cherokee Arrow 3
P28T	Piper	PA-28RT-201T Turbo Arrow 4
P28T	Piper	PA-28RT-201 Arrow 4
P32R	Piper	PA-32R-300 Cherokee Lance
P32R	Piper	PA-32R-300 Lance
P32R	Piper	PA-32R-301 Saratoga 2 HP
P32T	Piper	PA-32RT-300T Turbo Lance 2
P46T	Piper	PA-46-500TP Malibu Meridian
PA11	Piper	PA-11 Cub Special
PA12	Piper	PA-12 Super Cruiser
PA14	Piper	PA-14 Family Cruiser
PA15	Piper	PA-15 Vagabond
PA16	Piper	PA-16 Clipper
PA17	Piper	PA-17 Vagabond
PA18	Piper	PA-18 Super Cub
PA20	Piper	PA-20 Pacer
PA22	Piper	PA-22 Caribbean
PA22	Piper	PA-22 Colt
PA22	Piper	PA-22 Tri-Pacer

Type	Manufacturer & Model	
PA23	Piper	PA-23 Apache
PA24	Piper	PA-24 Comanche
PA25	Piper	PA-25 Pawnee
PA27	Piper	PA-23-250 Turbo Aztec
PA27	Piper	PA-23 Aztec
PA30	Piper	PA-30 Twin Comanche
PA31	Piper	PA-31
PA32	Piper	PA-32
PA34	Piper	PA-34 Seneca
PA36	Piper	PA-36 Pawnee Brave
PA38	Piper	PA-38 Tomahawk
PA44	Piper	PA-44 Seminole
PA46	Piper	PA-46-350P Malibu Mirage
PA46	Piper	PA-46-310P Malibu
PAT4	Piper	PA-31T3-500 T-1040
PAY1	Piper	PA-31T1-500 Cheyenne 1
PAY2	Piper	PA-31T-620 Cheyenne 2
PAY4	Piper	PA-42-1000 Cheyenne 400
PILL	Piper	PA-28R-300 Pillán
S108	Piper	108 Voyager
S108	Piper	108 Station Wagon
PTMS	Pitts / Aviat	S-12 Super Stinker
PTMS	Pitts / Aviat	S-12 Macho Stinker
PTS1	Pitts / Aviat	S-1 Special
PTS2	Pitts / Aviat	S-2 Special
PTSS	Pitts / Aviat	S-1-11 Super Stinker
I153	Polikarpov	I-153
I15B	Polikarpov	I-15bis
I16	Polikarpov	I-16
PO2	Polikarpov	Po-2
FOUG	Potez	CM-175 Zéphyr
FOUG	Potez	CM-170R Magister
PO60	Potez	600 Sauterelle
PO60	Potez	60 Sauterelle
ELST	Pützer	Elster
AN2	PZL Mielec	An-2 Antek
AN28	PZL Mielec	M-28 Bryza
AN28	PZL Mielec	An-28 Bryza
M15	PZL Mielec	M-15 Belphegor
M18	PZL Mielec	M-18 Dromader
M18T	PZL Mielec	M-18 Turbine Dromader
MG15	PZL Mielec	LiM-1
MG15	PZL Mielec	LiM-2
MG17	PZL Mielec	LiM-5
MG17	PZL Mielec	LiM-6

Type	Manufacturer & Model	
PA34	PZL Mielec	M-20
PO2	PZL Mielec	CSS-13
TS11	PZL Mielec	TS-11 Iskra
TS8	PZL Mielec	TS-8 Bies
PO2	PZL Okecie	CSS-13
PZ01	PZL Okecie	PZL-101 Gawron
PZ02	PZL Okecie	PZL-102 Kos
PZ04	PZL Okecie	PZL-104 Wilga
PZ05	PZL Okecie	PZL-105 Flaming
PZ06	PZL Okecie	PZL-106 Kruk
PZ12	PZL Okecie	PZL-112 Koliber Junior
PZ26	PZL Okecie	PZL-126 Mrówka
PZ3T	PZL Okecie	PZL-130 Orlik
PZ4M	PZL Okecie	PZL-104M Wilga 2000
PZ6T	PZL Okecie	PZL-106AT Turbo Kruk
RALL	PZL Okecie	PZL-110 Koliber
YK12	PZL Okecie	Yak-12
I23	PZL Swidnik	I-23 Manager
MI2	PZL Swidnik	Mi-2
PSW4	PZL Swidnik	SW-4
W3	PZL Swidnik	W-3 Sokól
F406	Reims	F406 Vigilant
F406	Reims	F406 Caravan 2
P337	Reims	FT337G Pressurized Skymaster
RC3	Republic	RC-3 Seabee
FANT	RFB	Fantrainer
D250	Robin / Apex	DR-200
D253	Robin / Apex	DR-253 Regent
DR10	Robin / Apex	Ambassadeur
HR10	Robin / Apex	HR-100
HR20	Robin / Apex	HR-200
R100	Robin / Apex	R-1180 Aiglon
R200	Robin / Apex	Alpha
R300	Robin / Apex	R-3120
R300	Robin / Apex	R-3140
R300	Robin / Apex	R-3100
R300	Robin / Apex	R-300
R300	Robin / Apex	R-3000
AC11	Rockwell	114 Commander 114
AC11	Rockwell	112 Commander 112
AC50	Rockwell	500 Shrike Commander
AC6L	Rockwell	685 Commander 685
AC90	Rockwell	690 Jetprop Commander 840
AC90	Rockwell	690 Turbo Commander 690
AC95	Rockwell	695 Jetprop Commander 980 / 1000

Type	Manufacturer & Model	
RC70	Rockwell	700 Commander 700
SS2P	Rockwell	Thrush Commander
V10	Rockwell	OV-10 Bronco
IR27	ROMBAC	IAR-827 Dacic
SE5R	Royal Aircraft Factory	SE-5A Replica
R90F	Ruschmeyer	R-90-230FG
R90R	Ruschmeyer	R-90-230RG
R90T	Ruschmeyer	R-90-420AT
NAVI	Ryan	Navion
PT22	Ryan	PT-22 Recruit
PT22	Ryan	ST-3KR Recruit
RYST	Ryan	ST-A / ST-M / ST-R
RYST	Ryan	PT-20
MF17	Saab	MFI-17 Supporter
MF17	Saab	MFI-15 Safari
SB05	Saab	105
SB91	Saab	91 Safir
G164	Schweizer	G-164 Ag-Cat
G64T	Schweizer	G-164 Turbo Ag-Cat
G64T	Schweizer	G-164 Ag-Cat Turbine
H269	Schweizer	269
S330	Schweizer	269D 330
S330	Schweizer	269D 333
SA37	Schweizer	SA-2-37A Condor
BDOG	Scottish Aviation	Bulldog
JS20	Scottish Aviation	Jetstream T.Mk.2
JS20	Scottish Aviation	Jetstream T.Mk.1
TPIN	Scottish Aviation	Twin Pioneer
MG17	Shenyang	J-5
MG17	Shenyang	F-5
SC7	Shorts	SC-7 Skyvan
TUCA	Shorts	Tucano
F260	SIAI-Marchetti	SF-260
F26T	SIAI-Marchetti	SF-260TP
F600	SIAI-Marchetti	SF-600 Canguro
FN33	SIAI-Marchetti	FN-333 Riviera
S05F	SIAI-Marchetti	S-205-18F / S-205-20F
S05R	SIAI-Marchetti	S-205-18R / S-205-20R / S-205-22R
S208	SIAI-Marchetti	S-208
S211	SIAI-Marchetti	M-311
S211	SIAI-Marchetti	S-211
SM19	SIAI-Marchetti	SM-1019
S223	SIAT	223 Flamingo
S38	Sikorsky	S-38 Replica
S39	Sikorsky	S-39

Type	Manufacturer	& Model
S51	Sikorsky	S-51
S51	Sikorsky	HO3S
S52	Sikorsky	HO5S
S52	Sikorsky	S-52
S55P	Sikorsky	S-55
S55P	Sikorsky	CH-19
S55P	Sikorsky	H-19 Chickasaw
S55P	Sikorsky	HH-19
S55T	Sikorsky	S-55T
S58P	Sikorsky	SH-34 Seabat
S58P	Sikorsky	HH-34 Seahorse
S58P	Sikorsky	CH-34 Chocktaw
S58P	Sikorsky	S-58
S58T	Sikorsky	S-58T
S62	Sikorsky	S-62
S62	Sikorsky	HH-52 Seaguard
S76	Sikorsky	S-76
SJ30	Sino Swearingen	SJ-30
NIPR	Slingsby	T-66 Nipper
RF6	Slingsby	Firefly
GA7	SOCATA	TB-360 Tangara
GY80	SOCATA	GY-80 Horizon
MS18	SOCATA	MS-200FG Morane
MS25	SOCATA	MS-200RG Morane
MS30	SOCATA	MS-300 Epsilon 2
ST10	SOCATA	ST-10
TAMP	SOCATA	TB-9 Tampico
TAMP	SOCATA	Tampico
TB30	SOCATA	TB-30 Epsilon
TB31	SOCATA	TB-31 Omega
TBM7	SOCATA	TBM-700
TBM8	SOCATA	TBM-850
TOBA	SOCATA	TB-10 Tobago
TRIN	SOCATA	TB-20 Trinidad
G2GL	SOKO	G-2 Galeb
G4SG	SOKO	G-4 Super Galeb
GAZL	SOKO	H-45 Partizan
JAST	SOKO	Jastreb
KRAG	SOKO	P-2 Kraguj
KRAG	SOKO	J-20 Kraguj
CAML	Sopwith	Camel
STRI	Sopwith	Triplane Replica
RF4	Sportavia-Pützer	RF-4
RF5	Sportavia-Pützer	RF-5
RS18	Sportavia-Pützer	RS-180 Sportsman

Type	Manufacturer & Model	
ST75	Stearman	PT-13 Kaydet
ST75	Stearman	Kaydet
ST75	Stearman	PT-17 Kaydet
ST75	Stearman	75 Kaydet
L5	Stinson	L-5 Sentinel
RELI	Stinson	Reliant
S10	Stinson	10 Voyager
S108	Stinson	108 Voyager
S108	Stinson	108 Station Wagon
SM60	Stinson	SM-6000 Tri-Motor
TRMA	Stinson	Tri-Motor (A)
SU26	Suchoi / Sukhoi	Su-26
SU29	Suchoi / Sukhoi	Su-29
SU31	Suchoi / Sukhoi	Su-31
SU38	Suchoi / Sukhoi	Su-38
ALO2	Sud	SA-318 / SA-3180 / SE-313 / SE-3130 Alou
ALO3	Sud	SA-316 / SA-319 / SE-3160 Alouette 3
ALO3	Sud	SA-319 Alouette 3
DJIN	Sud	SO-1221 Djinn
FOUG	Sud	CM-170R Magister
GAZL	Sud	SA-341 / SA-342 Gazelle
LAMA	Sud	SA-315 Lama
PUMA	Sud	SA-330 Puma
ALO2	Sud-Est	SE-3130 Alouette 2
DJIN	Sud-Ouest / SNCASO	SO-1221 Djinn
SASP	Supermarine	Spitfire Mk25
SPIT	Supermarine	Seafire
SPIT	Supermarine	Spitfire
SW2	Swearingen	SA-26 Merlin 2
SW3	Swearingen	Merlin 3
AUS3	Taylorcraft	Auster 3
AUS4	Taylorcraft	Auster 4
AUS5	Taylorcraft	Auster 5 Alpha
AUS6	Taylorcraft	Auster AOP6
J1	Taylorcraft	J-1 Autocrat
PLUS	Taylorcraft	Auster 1 / Plus C / Plus D
TA15	Taylorcraft	15 Foursome
TA15	Taylorcraft	15 Tourist
TA20	Taylorcraft	20 Zephyr 400
TA20	Taylorcraft	20 Seabird
TA20	Taylorcraft	20 Ranchwagon
TA20	Taylorcraft	20 Topper
TAYA	Taylorcraft	A
TAYB	Taylorcraft	BC
TAYD	Taylorcraft	DC

Type	Manufacturer & Model	
TF19	Taylorcraft	Sportsman (F-19)
TF21	Taylorcraft	F-21
TF22	Taylorcraft	F-22
I3	Technoavia (Interavia)	I-3
SG92	Technoavia (Interavia)	SM-92T Turbo Finist
SL90	Technoavia (Interavia)	I-1
SM20	Technoavia (Interavia)	SM-2000
SM92	Technoavia (Interavia)	SM-92 Finist
SP55	Technoavia (Interavia)	SP-55
SP91	Technoavia (Interavia)	SP-91 Slava
SP95	Technoavia (Interavia)	SP-95
Y18T	Technoavia (Interavia)	SM-94
Y18T	Technoavia (Interavia)	SM-2000P
AEST	Ted Smith	Aerostar
A660	Thrush	S-2R-T660 Turbo Thrush
SS2T	Thrush	Turbo Thrush (S-2RHG-T34/65)
B06T	Tridair	206L-ST Gemini ST
UU12	Udet	U-12 Flamingo Replica
UT60	UTVA	60
UT65	UTVA	65 Privrednik
UT66	UTVA	66
UT75	UTVA	75
RV10	Vans	RV-10
RV3	Vans	RV-3
RV4	Vans	RV-4
RV4T	Vans	RV-4T
RV6	Vans	RV-6
RV7	Vans	RV-7
RV8	Vans	RV-8
RV9	Vans	RV-9
VIMY	Vickers	FB-27 Vimy Replica
ACSR	Victa / AESL	Aircruiser
TOUR	Victa / AESL	Airtourer
VO10	Volaircraft	Volaire 10
VALI	Vultee	BT-13 Valiant
VALI	Vultee	BT-15 Valiant
WAC9	WACO	9
WACA	WACO	IBA, PBA, PLA, RBA, UBA, ULA
WACC	WACO	YKC, YKS, YOC, YQC, ZGC, ZKS, ZQC
WACC	WACO	CJC, CUC, DJC, DQC, EGC, EQC, QDC
WACC	WACO	UEC, UIC, UKC, UKS, VKS
WACC	WACO	AGC, AQC
WACD	WACO	S3HD
WACE	WACO	E (ARE/HRE/SRE) Aristocrat
WACF	WACO	CPF, ENF, INF, KNF, PBF, PCF, QCF

Type	Manufacturer & Model	
WACF	WACO	UBF, UMF, UPF
WACF	WACO	YMF, YPF, ZPF
WACF	WACO	RNF
WACG	WACO	CRG
WACM	WACO	JWM, JYM
WACN	WACO	ZVN
WACN	WACO	AVN
WACO	WACO	QSO
WACO	WACO	GXE
WACO	WACO	BSO, CSO, CTO, DSO
WACO	WACO	ASO, ATO
WACO	WACO	125
WACO	WACO	PSO
WACO	WACO	10
WACT	WACO	RPT
WACF	WACO Classic	YMF
D11	Wassmer	D-120 Paris-Nice
D11	Wassmer	D-112
WA40	Wassmer	WA-40 Super 4
WA41	Wassmer	WA-41 Baladou
WA42	Wassmer	WA-421 Prestige
WA50	Wassmer	WA-51 Pacific
WA50	Wassmer	WA-54 Atlantic
WA50	Wassmer	WA-52 Europa
WA80	Wassmer	WA-80 Piranha
B47G	Westland	Sioux
LYNX	Westland	Lynx
LYNX	Westland	Super Lynx
LYSA	Westland	Lysander
PUMA	Westland	Puma
S55T	Westland	WS-55 Whirlwind 3
SCOU	Westland	Scout
WASP	Westland	Wasp
WESX	Westland	Wessex
WG30	Westland	30
Y112	Yakovlev / Jakovlev	Yak-112
Y18T	Yakovlev / Jakovlev	Yak-18T
YAK3	Yakovlev / Jakovlev	Yak-3
YAK9	Yakovlev / Jakovlev	Yak-9
YK11	Yakovlev / Jakovlev	Yak-11
YK12	Yakovlev / Jakovlev	Yak-12
YK18	Yakovlev / Jakovlev	Yak-18
YK18	Yakovlev / Jakovlev	Yak-18U
YK50	Yakovlev / Jakovlev	Yak-50
YK52	Yakovlev / Jakovlev	Yak-52

Type	Manufacturer & Model	
YK53	Yakovlev / Jakovlev	Yak-53
YK54	Yakovlev / Jakovlev	Yak-54
YK55	Yakovlev / Jakovlev	Yak-55
YK58	Yakovlev / Jakovlev	Yak-58
EDGE	Zivko	Edge 540
BU81	Zlin	Z-183
BU81	Zlin	Z-181
BU81	Zlin	Z-182
SAVG	Zlin	Savage
Z22	Zlin	Zlin Z-22 Junak
Z22	Zlin	Z-22 Junak
Z26	Zlin	Zlin Z-226 Akrobat
Z26	Zlin	Zlin Z-526 Akrobat
Z26	Zlin	Zlin Z-326 Trener Master
Z26	Zlin	Zlin Z-26 Trener
Z26	Zlin	Zlin Z-326 Akrobat
Z26	Zlin	Zlin Z-626
Z26	Zlin	Zlin Z-526 Trener Master
Z26	Zlin	Z-26 Trener
Z26	Zlin	Zlin Z-726 Universal
Z26	Zlin	Z-226 Akrobat
Z26	Zlin	Z-126 Trener 2
Z26	Zlin	Z-326 Trener Master
Z26	Zlin	Z-626
Z26	Zlin	Zlin Z-226 Trener 6
Z26	Zlin	Zlin Z-126 Trener 2
Z26	Zlin	Z-526 Akrobat
Z26	Zlin	Z-326 Akrobat
Z26	Zlin	Z-226 Trener 6
Z26	Zlin	Z-726 Universal
Z26	Zlin	Z-526 Trener Master
Z37P	Zlin	Z-37 Cmelák
Z37P	Zlin	Z-237 Cmelák
Z37T	Zlin	Z-137 Agro Turbo
Z37T	Zlin	Z-37T Agro Turbo
Z42	Zlin	Zlin Z-142
Z42	Zlin	Z-142
Z42	Zlin	Z-42
Z42	Zlin	Zlin Z-42
Z42	Zlin	Z-242
Z42	Zlin	Zlin Z-242
Z43	Zlin	Zlin Z-143
Z43	Zlin	Z-43
Z43	Zlin	Z-143
Z43	Zlin	Zlin Z-43

Mnemonics

Aviation Mnemonics

Here are some handy mnemonics that pilot use to help verify
the completion of certain tasks. Of course, conscientious
checklist usage should be primary, but these can also help.

CIGAR	Quick Runup/Ground Check
C	controls check
I	instruments set
G	gas (proper tank, pump on, etc)
A	aircraft configuration (flaps, trim, etc.)
R	runup

TOMATO FLAMES	(VFR) Day Minimum Equipment 91.205b
T	tachometer
O	oil temperature
M	magnetic compass
A	altitude
T	temp gauge if liquid cooled
O	oil pressure
F	fuel gauge
L	landing gear position light
A	airspeed indicator
M	manifold pressure
E	ELT
S	Shoulder harness, if certified after 7/18/78.
F	Floatation device, if for hire, beyond the power-off gliding distance of the shoreline
A	Anti-collision strobe, if certified after 3/11/96.

FLAPS	(VFR) Night Add'l Requirements 91.205c
F	Fuses, set of 3 each type
L	Landing light if for hire
A	Anti collision strobe, if certified after 8/11/71.
P	position lights
S	source of electric (adequate)

AV1ATE	Required Inspections
A	Annual Inspection
V	VOR Test, 30 days
1	100 hour, if for hire
A	Altimeter/Static, 24 month
T	Transponder, 24 month
E	ELT, 12 months & ELT Battery, 1/2 battery life or 1 hour of continuous use

ARROW	Required Documents
A	Airworthiness Certificate
R	Radio Telephone License (International flights)
R	Registration, current
O	Operating limitations (POH), Markings & Placards
W	Weight & Balance data

CAPER	Maneuvers Checklist
C	Clearing Turns
A	Altitude, proper for maneuver
P	Proper power & entry to maneuver
E	Execute maneuver
R	Recover from maneuver

Darren Smith

LCA	Line-Up Check
LIGHTS	strobes, navs, landing
CAMERA	transponder (so ATC can "see" you)
ACTION	any other actions to be performed like boost pump on, control checks, flaps and trim set, etc.

BLITTS	Line-Up Check
B	boost pump on
L	lights as required
I	instruments set
T	transponder on
T	takeoff time noted
S	seat, belts, doors secured

GUMPS	Before Landing
G	gas (proper tank, pumps, mixture rich, etc.)
U	Undercarriage
M	mixture set
P	prop set and/or primer in/locked
S	switches (lights, pitot heat, etc.) & "see" gear
Note: add C to GUMPS for carb heat (becomes "Charlie GUMPS")	

MFACTS	After Landing Checklist
M	Mixture set
F	Flaps Up
A	Aux Fuel Pumps Off
C	Cowl Flaps
T	Trim & Transponder, standby
S	Switches (lights, pumps, pitot heat, strobes if night)

Aviation Acronyms, Contractions, & Mnemonics (PocketLearning)

MIDGET	Shutdown
M	master off
I	ignition off
D	doors/windows locked
G	gust lock installed
E	ELT off
T	tiedown plane

Simple Shutdown	
1st	Avionics – Mixture – Master – Mags
2nd	Gust Lock – Doors –Tiedown

FLARE	Enroute
F	flaps set (if extended during takeoff)
L	lights as required
A	auxiliary fuel pump off (if on for departure)
R	radar transponder on
E	engine (lean mixture when at altitude)

HASELL	Aerobatic Maneuvers Checklist
H	Height - sufficient to recover
A	Airframe - rated for the maneuver, flaps and landing gear as required, trimmed
S	Security - hatches and harnesses secure, no loose items in cockpit, gyros caged
E	Engine - running normally, fuel sufficient for maneuver, no carb icing
L	Location - clear of cloud, controlled airspace, airfields, built-up areas, other aircraft
L	Look Out - inspection turn to make sure area is clear around and below

I'M SAFE	Personal Preflight Checklist
I	Illness
M	Medications
S	Stress
A	Alcohol
F	Fatigue
E	Emotion –or– Eating –or– Environment

SETVODA	Class E Airspace - the 7 floors of Class E
S	Surface Area - dashed magenta line
E	Extension Area (to class B, C, D)
T	Transition (beginning at either 700' or 1200' agl)
V	Victor airways (1200' agl up to 17,999msl)
O	Off shore (12NM -usually designated by blue zipper)
D	Domestic enroute airspace
A	Above (above class G, D, C, B, A, above 14,500msl)

PARE	Spin Recovery
P	Power - idle
A	Ailerons - neutral
R	Rudder - opposite direction of rotation
E	Elevator - down(forward) to break stall once spin/ rotation has stopped or under control

ABCD	Single Engine Out - Emergency
A	Airspeed - best glide
B	Best field or landing surface
C	Communicate (above 2000') & Checklist (below 2000')
D	Declare emergency

E-BGAAR	Aircraft Right of Way Rules
E	Emergencies
B	Balloon
G	Glider
A	Airship
A	Aircraft
R	Rotorcraft

WARM PC	Special Use Airspace
W	Warning Area
A	Alert Area
R	Restricted Area
M	MOA - Military Operations Area
P	Prohibited Area
C	Controlled Firing Area

NWA KRAFT*	Questions to ask your student about his preflight
N	NOTAMs including TFRs
W	Weather (AWOS, ATIS, ASOS, etc.)
A	ATC Delays
K	Killer items (Gust lock, oil & fuel level)
R	Runway: length, selection, performance
A	Alternate if flight can not be completed
F	Fuel: quantity, performance, regulatory requirements
T	Takeoff & Landing: performance

* Now that the airline NWA is gone, perhaps some-
one will come up with something more significant.

CRAFT	IFR Clearance or Tower VFR Departure
C	Cleared to (location, airport, etc)
R	Routing (initial heading, or full routing, "as filed", etc)
A	Altitude (initial, and expected)
F	Frequency (departure)
T	Transponder code

TTTTTT	(IFR) Crossing a Fix / Approach
T	turn to proper heading
T	time hold or approach
T	twist OBS knob to inbound course
T	throttle adjustments, as required
T	talk - procedure turn inbound, entering the hold, etc.
T	Track inbound
T	Tires, GUMPS check

WIRE-TAP	(IFR) Nearing Destination
W	Weather (AWOS, ATIS, ASOS, etc.)
I	instruments set
R	radios tuned
E	elevation (check final approach fix altitude)
T	talk to ATC
A	altitudes for decision height or minimum descent altitude
P	procedure for missed approach

Aviation Acronyms, Contractions, & Mnemonics (PocketLearning)

HAMSACC	(IFR) Required Reported Items
H	holding (time and altitude)
A	altitude changes
M	missed approach
S	safety of flight (if anything affects it)
A	airspeed changes (of 5% or 10kts)
C	communication or navigation capability loss
C	climb rate (when unable to maintain 500fpm)

GRAB CARD	(IFR) Required Equipment 91.205d
G	Generator/Alternator
R	Radios, appropriate for flight
A	Altimeter, sensitive
B	Ball, Slip/Skid Indicator
C	Clock/Timer
A	Attitude Indicator
R	Rate of Turn Indicator
D	Directional Gyro

SWAPAFO	Briefing of Threats to Flight Safety
S	Security
W	Weather
A	Airport
P	Pilot
A	Aircraft
F	Flight Plan
O	Other

TIVO	Anytime you use a NAVAID
T	Tune
I	Identify (morse code)
V	Verify (frequencies, charts, etc)
O	Orient (yourself with the NAVAID and on the chart)

ANDS	Compass Dip Error (northern hemisphere)
A	Accelerate (on an easterly or westerly heading)
N	North (is depicted in error - momentarily)

D	Decelerate (on an easterly or westerly heading)
S	South (is depicted in error - momentarily)

UNOS	Compass Turning Error (northern hemisphere)
U	Undershoot (because of lag in compass)
N	North (if on N heading, undershoot turns to the E or W by 30°)

O	Overshoot (because of lead in compass)
S	South (if on S heading, overshoot turns to the E or W by 30°)

TURNPALE	Aircraft Airworthiness Certificate
T	Transport
U	Utility
R	Restricted
N	Normal
P	Provisional type
A	Aerobatic
L	Limited (typically surplus military aircraft in civilian use)
E	Experimental (typically homebuilt or exhibition)

"Real knowledge is to know the extent of one's ignorance."
— *Confucius*

Appendix

Personal Minimums Checklist

ADAPTATION REPRINT OF FAA P-8740-55 AFS-810(1996)

PILOT

AIRCRAFT

EN**V**IRONMENT

EXTERNAL
PRESSURES

Your Personal Minimums Checklist
○ An easy-to-use, personal tool, tailored to your level of skill, knowledge & ability.
○ Helps you control and manage risk by identifying even subtle risk factors
○ Lets you fly with less stress and less risk.

Practice "Conservatism Without Guilt"
Each item provides you with either a space to complete a personal minimum or a checklist item to think about. Spend some quiet time completing each blank & consider other items that apply to your personal minimums. Give yourself permission to choose higher minimums than those specified in the regulations, aircraft flight manuals, or other rules.

How to Use Your Checklist
Use this checklist just as you would one for your aircraft. Carry the checklist in your flight kit. Use it at home as you start planning a flight and again just before you make your final decision to fly. Photocopy the next page (front & back) and carry extras in your flight bag. Be wary if you have an item that's marginal in any single risk factor category. But if you have items in more than one category, you may be headed for trouble.

If you have marginal items in two or more risk factors/ categories, don't go!
Periodically review and revise your personal minimums checklist as your personal circumstances change, such as your proficiency, recency, or training. You should never make your minimums less restrictive unless a significant positive event has occurred. However, it is okay to make your minimums more restrictive at any time. Never make your minimums less restrictive when you are planning a specific flight, or else external pressures will influence you.

PILOT

Experience/Recency

Takeoffs/Landings ____In the last
____days

Hours in make/model ____In the last
____days

Instrument approaches ____In the last
(simulated or actual) ____days

Instrument flight hours ____In the last
(simulated or actual) ____days

Terrain and airspace ____Familiar?

Physical Condition

Illnesses, none in the last ____days

Medication/Drugs, none in ____days

Stressful Event, none in ____days

Alcohol, none in the last ____In the last 24hrs

Fatigue: hours of sleep ____In the last 24hrs

Eating/Nourishment/Water ____hours ago

_____

_____

_____

_____

Thanks to:

FAA Aviation Safety Program
The Ohio State University
King Schools

IRCRAFT

Fuel Reserves

VFR Day ____hours

Night ____hours

IFR Day ____hours

Night ____hours

Experience in type

Takeoffs/Landings ____In the last

(in aircraft type) ____days

Aircraft Performance

Consider the following:

- Gross weight ____
- Load distribution ____
- Density Altitude ____
- Performance Charts ____

Ensure you have a margin of safety

Aircraft Equipment

Avionics/GPS, familiar with ____

Autopilot, familiar with ____

COM/NAV, appropriate ____

Charts, current & adequate ____

Clothing, suitable for flight ____

Survival gear, suitable for flight ____

Required Documents (ARROW) ____

Required Inspections (AVIATE) ____

Required Equipment (§91.205) ____

Other ____

ENIRONMENT

Airport Conditions

Crosswind, Departure   _____% max POH

Crosswind, Arrival   _____% max POH

Runway length, Departure_____% over POH

Runway length, Arrival   _____% over POH

Weather

Forecast, not more than   _____Hours old

Icing conditions, familiar   _____

Weather for VFR

Ceiling	Day	_____feet
	Night	_____feet
Visibility	Day	_____miles
	Night	_____miles

Weather for IFR

Precision Approaches

Ceiling   _____ft above min

Visibility   _____mi above min

Non-Precision Approaches

Ceiling   _____ft above min

Visibility   _____mi above min

Missed Approaches

No more than   _____before divert

Takeoff Minimums

Ceiling   _____feet

Visibility   _____miles

_____

_____

_____

_____

_____

_____

EXTERNAL PRESSURES

Trip Planning
Allowance for delays, _____ minutes

Diversion/Cancellation Alternate Plans
✓ Notification of person(s) you are meeting.
✓ Passengers briefed on diversion/cancellation plans and alternates.
✓ Modification or cancellation of car rental, restaurant, or hotel reservations.
✓ Alternate transportation (air/car/etc)

Personal Equipment
✓ Credit card & telephone numbers available for alternate plans.
✓ Appropriate clothing or personal needs (eyewear, medication…) in the event of unexpected stay.

Importance of Trip
The more important the trip, the more tendency there is to compromise your personal minimums, and the more important it becomes to have alternate plans.

For More Information, Contact:

Darren Smith, ATP, CFII/MEI
Certificated Flight Instructor
www.cfidarren.com

7-Day Instrument Rating Training

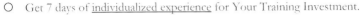

Structured, Efficient, & Cost Effective:

○ Get 7 days of <u>individualized experience</u> for Your Training Investment.

○ <u>Gain real confidence</u> working the IFR system with airline style routes.

○ <u>Learn real world techniques</u> from an experienced Instructor (CFII)

○ Build IFR skills in your aircraft - in a format convenient to your schedule.

Want to finish your IFR Rating in 7 days?

If you're interested in getting your instrument in 7 days, this is the program for you. There are no gimmicks to this, it's hard work, and requires your full concentration. But when you're finished, you'll have gained solid experience in the real world of IFR.

Don't spend 9-12 months and $7,000 to $10,000 on your instrument rating! Condensed training experiences will not only save you money, but the intensity guarantees that the lessons last a lifetime. Your training is accelerated and less time is spent reviewing the last lesson... a real money saver. None of the other accelerated programs provide the comprehensive, high quality learning materials: the Oral Exam Prep Kit and the Instrument Rating Checkride Reviewer.

The three-day IFR Adventure is an integral part of the seven day IFR Rating and guarantees a unique training experience. We integrate the IFR Adventure into the course not only to meet certain instrument rating requirements, but also to serve as a real world capstone experience which increases your confidence in the IFR system.

What if you want a slower pace than a 7 day instrument rating training course? A 10-day program is available at a slightly higher cost.

Key Features:

○ The program includes up to 56 hours of one-on-one ground & flight instruction.

○ You will receive <u>at least</u> 30 hours of one-on-one flight instruction on some of the most challenging instrument approaches in the South East. You'll fly at least 25 approaches during the program, many with SID's STAR's and holding patterns. You'll be astonished at how much you will learn.

○ When you are not in the air, you will learn techniques that will simplify otherwise complex procedures such as procedure turns, holding pattern entries and many others.

○ Includes a 3 day IFR Adventure which adds up to a true flying adventure that will make you a more knowledgeable and confident pilot... plus it's a whole lot of fun!

○ You are PIC for the entire program receiving PIC credit in your logbook for 100% of the flight time.

○ We know your aircraft—we've got time virtually all aircraft & helicopters and avionics configurations.

6 Reasons to Choose This Program:

1. This is a quick, efficient, structured training program to get you through this training with minimum pain, maximum quality.
2. We constantly evaluate our competitors to ensure that our program provides the highest quality at the most reasonable cost in the country... compare for yourself. We're not beat by anyone.
3. You are PIC for the entire program receiving PIC credit in your logbook for 100% of the flight time.
4. Safety: techniques for safe instrument flight are emphasized by your instructor.
5. Personal attention is provided as you progress through your training program. You'll never wonder about your progress. You will never fall through the cracks .
6. The chief instructor has been a teacher since 1996, is a Master Instructor (NAFI), Advanced Ground Instructor rated (FAA), a Gold Seal Certificated Flight Instructor (FAA), an Aviation Safety Counselor (FAA), and a former airline pilot. Your instructor is a highly qualified ATP-rated, FAA Gold Seal, NAFI Master instructor with 100% pass rate. None of the other local providers use an instructor of this calibre.

Prerequisites for Instrument Rating:

You'll be ready for a checkride immediately upon completion if you have:
O   a passing score on your instrument written,
O   at least 50 hours PIC X/C & at least 40 hours of instrument time,
O   and meet PTS requirements

Those that don't meet these requirements will receive instruction at reduced rate. This can be completed before or after the trip. What if you want a slower pace than a 7 day instrument rating training course?   A 10-day program is available at a slightly higher cost.

Availability:

Call for scheduling.   Includes: All flight and ground instruction, over 56 hours combined. Orientation ground package.

Not Included: Aircraft or anything related to the aircraft (Fuel, Oil, Insurance, Rental, Gov Fees, Repair, Etc), Meals, Ground Transportation & Lodging for you & your instructor, entertainment or any personal expenses you might encounter. You must supply your own aircraft, oil & fuel.

Participant Comments:

"You have a 'teaching way' about you and its very effective." Dave C
"You really have this down to a science." Alan D
"Thank you for the many flights of stressing good flying habits.
Its now paying off in my Air Force career." Mark S
"My training was second to none. My skills are consistently
complimented by other flight instructors I have flown with" Tim N

www.ifrnow.com

South East IFR Adventure

In Your Aircraft

Our Practice Area is a 2,000 mile Cross Country Training Adventure!

Intense 3-day IFR Confidence Builder

- Get More Individualized Experience For Your Training Investment.
- Gain real confidence working the IFR system with airline style routes.
- Learn Real World Techniques from an experienced Instructor (CFII)
- Finish your instrument rating or IPC & build IFR skills in your aircraft.
- We can start your IFR adventure anyplace and customize it to your location.
- Get more than current, get proficient.

Want a real IFR workout?

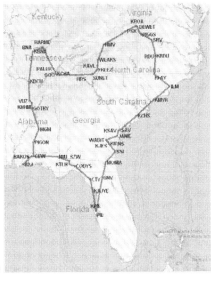

If you're interested in finishing your instrument rating or gaining solid experience in the real world of IFR, here's a unique opportunity for an accelerated program for aircraft owners that is completed in **3 days**.

This Southeast Trip is a flying adventure, traveling up the east coast, down the Blue Ridge Mountain chain to the Great Smoky Mountain, through the heartland, and back down through the Mississippi Valley. You will gain the experience of flying over 2,000 miles and over 20 flight hours. Upon our return, you'll receive your IPC or endorsement for the check-ride. In addition, you'll enjoy spectacular scenery, gain valuable experience by training away from your local airport and receive professional instruction in a variety of flight conditions.

This training experience gives you the confidence you need to really use the IFR system. You'll operate in the environment like a professional by focusing on the practical aspects of instrument flight. You will also get a chance to really learn some of the most challenging instrument approaches in the country, performing them like the professionals.

Key Features:

- Your orientation includes ground instruction, briefings, flight planning.
- You will receive over 20 hours of one-on-one flight instruction on some of the most challenging instrument approaches in the Southeast.
- When you are not in the air, you will learn techniques that will simplify otherwise complex procedures such as procedure turns, holding pattern entries and many others.
- You'll receive over 10 hours of ground instruction during the course of the program... you will be confident you can pass your checkride.
- If instrument rated, you will receive your IPC immediately upon return. Instrument rating completion students will typically take their instrument checkride immediately upon returning.
- In all, you complete **at least** 10 approaches, many with SID's, STAR's, & holding.
- You'll be astonished at how much you will learn in this total immersion experience.
- Aircraft owners: this is the ultimate insurance checkout if you need it.
- All of this adds up to a true flying adventure that will make you a more knowledgeable and confident pilot...plus it's a whole lot of fun!
- We know your aircraft—we've got time virtually all aircraft & helicopters and avionics configurations.

Prerequisites for Instrument Rating:

You'll be ready for an IPC or IFR check-ride immediately upon return if you have:

- a passing score on your instrument written,
- at least 50 hours PIC X/C & at least 40 hours of instrument time,
- and meet PTS requirements

Those that don't meet these requirements will receive instruction at reduced rate. This can be completed before or after the trip.

Availability:

This program starts every other Friday. Call for scheduling. Other routes and schedules are available for this highly individualized program.

Includes: All flight and ground instruction, over 35 hours combined. Orientation ground package including Checkride Review Guide

Not Included: Aircraft or anything related to the aircraft (Fuel, Oil, Insurance, Rental, Gov Fees, Repair, Etc), Meals, Ground Transportation & Lodging for you & your instructor, entertainment or any personal expenses you might encounter. You must supply your own aircraft, oil & fuel.

www.ifradventure.com

Introducing…

One of the most comprehensive Flight Training websites available

Resources for all pilots:

- Articles on current practices and techniques to help you fly safely (in the Reading section)
- Links to other websites.
- Download section.
- FAA Safety Publications, Manuals, Practical Test Standards, and Training Guides.
- Free question & answer on difficult aviation topics, answered by an expert.
- An online store to purchase training products for your rating.
- Information on aviation ground schools, including syllabus, schedule, and enrollment info.
- Information resources for getting back into flying, including information & structure of Flight Reviews (BFR) and Instrument Proficiency Checks (IPC).
- An extensive graphical & textual weather products page with thumbnail views of all the current weather.
- Human Factors for pro pilots

Resources for helicopter pilots:

- Complete syllabus online including a maneuvers checklist.
- Reading library, articles, and information on the various helicopter ratings.
- Helicopter Lesson Plans for helicopter students.

Resources for multi-engine pilots:

- Complete syllabus online including a maneuvers checklist.

www.cfidarren.com

Responsibility & Authority of the PIC (§91.3)

The PIC is directly responsible for and the final authority in determining the airworthiness and operation of the aircraft. The PIC may deviate from any FAR to meet the requirements of an emergency. If the PIC deviates from and FAR, he or she shall, if requested, send a written report of the deviation to the Administrator.

Other Products by Darren Smith

Visi-Hold™ - *Know Instantly*

Never wonder about holding entries again! Works with standard holds and non-standard holds (left turns). The Visi-Hold™ package comes complete with Visi-Hold™ template, directions, and two articles on holds: All About Holding and Holding Simplified. You'll pay no less than $14.95 for a complicated sliderule holding pattern calculator made by ASA. That's not even including shipping! Those of you who know what real IFR is like know that you can't fool around with a sliderule while the airplane is bouncing around. Instead, get the original Visi-Hold™, helping pilots know the holding pattern entry since 2000. Price: $10

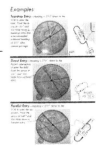

Guaranteed Pass Helicopter Flash Cards

This package includes 240 easy to read, soft yellow cards -- not flimsy paper -- which will help you pass your FAA written and oral (checkride) helicopter exams. Card size approximately: 4.25" x 2.75" - a stack of cards approximately 2" high. Includes all helicopter subjects. Price: $25

Visi-Plotter™ - Simply the Best Flight Planning Plotter Made

This is the only VFR plotter invented by a CFI for pilots. It includes innovative features that provides temperature conversion, flight planning scratch pad area, and a no-mistakes course protractor that even beginners won't be confused by. It provides scales for sectional & WAC charts, and it even has inches & MM measurements. Price $15

The Most Accurate

The Visi-Plotter navigational plotter is created from a calibrated hot stamp. It is the most accurate way to create a plotter. No other plotter on the market uses this production method because its more expensive than the silk screen method used by every other manufacturer.

Pilot's Rules of Thumb™ - the ultimate VFR/IFR Checklist

The ultimate checklist that every pilot needs. Ten years in development, this is a product that pilots of all levels of experience can immediately use. This tool covers 30 normal and non-normal situations and has 9 special tools that you'll use on every flight. This high-quality plastic ruler/checklist measures approximately 3.75" x 9". Click the picture to see a larger picture of this tool or click the link for more information. Price: $4

Instrument Rating Checkride Reviewer

Finally, instrument rating help! The best seller review guide is designed to help you to get through the Instrument Checkride. Includes a special offer for Visi-Hold™ (see other side) as a bonus. This package can be used as a self study guide or by flight instructors to provide IPC/Checkride preparation. Designed to fit in your flight bag, size: 6" x 9" Price: $25

Getting the Most from Your Flight Training

This is the essential guide to becoming a better pilot, paying the least, getting the most, and finishing as quickly as possible. Includes sections on Becoming a Better Pilot, Ground & Flight Instruction Tips, and Earning Certificates & Ratings. Designed to fit in your flight bag, size: 6" x 9" Price: v$15

Pilot's Radio Communications Guide

This review guide is designed as a quick reference guide or radio communications training self study guide. This covers VFR & IFR radio communications. Build your confidence by knowing what to say and when. Designed to fit in your flight bag, size: 6" x 9" Price: $15

Learning IFR Enroute Charts

This 40 page guide helps the Flight Instructor or Instrument Students with IFR Enroute Symbology. It uses a unique IFR Chart Extract to teach symbology. It includes all symbology you need to know for your Instrument checkride or Instrument Proficiency Check. As a bonus chapter, Instrument Approach Procedure (plate) symbology is also included. All of which you need to know prior to your checkride. Price: $10

I've Screwed Up! - Now What?

Practical advice and tips using the NASA form when you've violated the FARs. This 48 page guide will walk you step by step through filing a NASA safety report. It will tell you the techniques and traps to submit a good report AND keep your pilot certificate unblemished. Price $15.

Safer Approaches

Safer Approaches will teach you how to conduct Instrument Approach Procedures to a higher standard of safety and precision. You will learn:
- Four Fundamentals of Safe Approaches,
- How to virtually eliminate possibility of Controlled Flight Into Terrain,
- How to perform a Constant Angle Non-Precision Approach (CANPA),
- How to calculate a Visual Descent Point (VDP),
- How to practice building your flying precision.
Includes the Stabilized Approach Descent Rate Table, a plastic, kneeboard sized IFR tool that will eliminate the mental math applying these techniques during your IFR flying. 14pp. Price: $8

Winter Flying (PocketLearning)

Quick tips & techniques so you can get the most out of your winter flying - safely. Includes information on Winter Preflight, Induction Icing, Carb Icing, Airframe Icing, Tail Stalls, Hypothermia Strategies for Reducing the Risk, Winter Survival Kit, Aeronautical Decision Making, and the Personal Minimums Checklist. Size: 5" x 8" Price: $8

Online Ordering - Free Shipping
www.cfidarren.com

Made in the USA
Charleston, SC
27 June 2016